RADIOISOTOPES

The INTRODUCTION TO BIOTECHNIQUES series

Editors:

J.M. Graham MIC Medical Ltd, Merseyside Innovation Centre, 131 Mount Pleasant, Liverpool L3 5TF

D. Billington School of Biomolecular Sciences, Liverpool John Moores University, Byrom Street, Liverpool L3 3AF

CENTRIFUGATION

RADIOISOTOPES

Forthcoming titles

LIGHT MICROSCOPY

ANIMAL CELL CULTURE

GEL ELECTROPHORESIS: PROTEINS

RADIOISOTOPES

D. Billington

School of Biomolecular Sciences, Liverpool John Moores University, Byrom Street, Liverpool L3 3AF, UK

G.G. Jayson

School of Chemical and Physical Sciences, Liverpool John Moores University, Byrom Street, Liverpool L3 3AF, UK

and

P.J. Maltby

Department of Nuclear Medicine, Royal Liverpool Hospital, Prescot Street, Liverpool L7 8XP, UK

In association with the Biochemical Society

First published in the United Kingdom 1992 by
BIOS Scientific Publishers Limited,
St Thomas House, Becket Street, Oxford OX1 1SJ

A CIP catalogue for this book is available from the British Library.

ISBN 1 872748 85 6

Typeset by Enset Photosetting Ltd, Bath, UK
Printed by The Alden Press Ltd, Oxford, UK

Preface

Radioisotopes are used in all disciplines of the biological sciences. The main advantage of using radioisotopes is the exquisite sensitivity they give to biological experiments; when a substance is radiolabeled, it is possible to detect amounts as small as a few hundred molecules. Such sensitivity is also made use of in nuclear medicine, a rapidly expanding discipline of medicine, where radioisotopes are administered to man for either diagnostic or therapeutic reasons.

The basic aim of this book is to allow the biological scientist to design his/her own experiments using radioisotopes, and to carry them out safely. The book is aimed primarily at those new to working with radioisotopes, although it should be useful to those with some experience who wish to design new experiments. It sets out to explain what radioactivity is, how it can be introduced into biological molecules, and how it can be detected and measured. In addition, a brief introduction to the use of radioisotopes in nuclear medicine is given. Whilst this book is intended mainly for second and final year undergraduates, postgraduates, research technicians and academics in the biological sciences, it is hoped that it will also provide a useful introductory text for radiopharmacists and for medical students.

Part 1 of the book deals with basic principles and methods and commences with an introduction to radioactivity from the standpoint of atomic and nuclear structure. Modes and kinetics of radioactive decay are discussed, with particular reference to those radioisotopes commonly used in the biological sciences. Although most researchers buy radiolabeled chemicals from commercial suppliers, their methods of production are covered in order to remove the 'black box' approach to their origin. Methods of measurement of radioactivity are discussed and include liquid scintillation counting, sodium iodide scintillation counting, autoradiography and medical imaging.

It is clearly impossible in a book of this size to cover all of the applications of radioisotopes in the biological sciences. However, Part 2 of the book details some of these and concentrates on modern applications. It is divided into four chapters dealing with radioimmunoassay and ligand binding assays, tracer techniques, molecular biology and nuclear medicine.

In most, but not all, biological experiments, the amount of radioactivity used is small, which in turn presents only a small radiation hazard. Never-the-less, safety is of prime importance when working with radioisotopes and safe experimental practice is stressed throughout the book. Designated sections on radiation dosimetry, radiation protection and measurement in radiation safety are included in Part 1.

D. Billington
G.G. Jayson
P.J. Maltby

Contents

PART 1: BASIC PRINCIPLES AND METHODS

1 Introduction to Radioactivity

1.1 General atomic structure

The physical and chemical properties of an element depend upon the structure of its atoms which in turn depends upon the content of 'fundamental' particles. The Rutherford–Bohr model of the atom shows the heavy protons (p^+) and neutrons (n) concentrated in the small volume of the nucleus (diameter approximately 10^{-15} m) in the center of the atom, whilst the light electrons (e^-) surround and neutralize the nucleus at distances of up to 10^{-10} m (see *Figure 1.1* and *Table 1.1*). Within the atomic diameter, the electrostatic attraction of the electrons towards the protons of the nucleus, their repulsion away from each other and their spin, result in the electrons occupying a specific number of most probable (quantized) distance–energy positions from the nucleus. These positions are known as orbitals or shells and leave most of the atom empty.

1.2 The atomic nucleus

Unlike the atom (*Figure 1.1*), there is no precise anatomical 'picture' of the nucleus of the atom. Rather there are models which fit specific experimental results. The nearest we can get to visualizing the nucleus is as a liquid drop with an outer shell-like structure. In this model, the density of the nucleus is very much greater than that of the atom.

Nuclear particles are collectively known as nucleons. The major particles of the nucleus are protons and neutrons although the hydrogen atom is unique among the elements in that it does not possess any neutrons. Even these minute entities have sub-structures of quarks and gluons. In addition to the long-lived protons and neutrons, there are a large number (~100) of short-lived nucleons with half-lives of between 10^{-24} and 10^{-6} sec

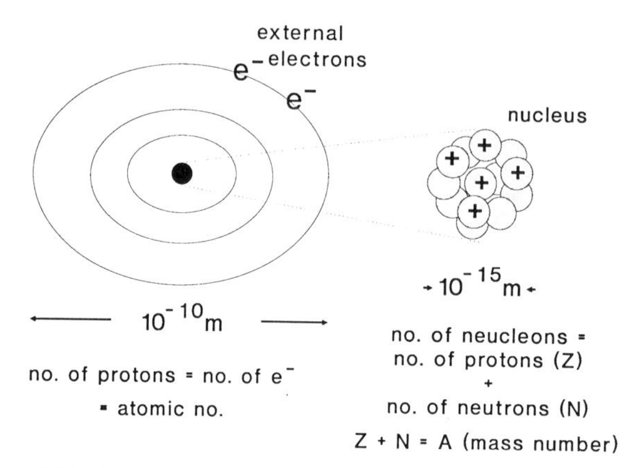

***FIGURE 1.1:** Model of the atom.*

(e.g. π- and κ-mesons, λ and ω particles, etc.) which appear momentarily during nuclear reactions. The mass, energy equivalence and electrostatic charge of the electron and the long-lived nucleons are shown in *Table 1.1.*

Despite the enormous repulsion between the positively charged protons over such small distances in the nucleus (10^{-15} m), they are held together with the neutrons by even stronger, short range, attractive nuclear forces. The binding energy for these strong attractive forces comes from the conversion of some of the masses of the nucleons into energy. This is often

***TABLE 1.1:** Physical properties of long-lived 'fundamental' particles in the atom and of ionizing radiations*

	Mass		Energy	
Particle	**10^{-27} kg**	**a.m.u.**	**(MeV)**	**Charge**
Fundamental particles				
Electron, e^-,e^+	0.000911	0.000549	0.51	−1 or +1
Proton, p^+	1.672079	1.007277	938.0	+1
Neutron, n	1.674385	1.008665	939.0	0
Neutrino, v	0			0
Ionizing radiations from radioactive isotopes				
Alpha, α^{2+}	6.642501	4.001507		+2
Beta, β^-,β^+	0.000911	0.000549	0.51	−1 or +1
Gamma-ray, γ	0			0

The standard atomic mass unit (a.m.u.) is taken as 1/12th of the mass of a ^{12}C atom. One a.m.u. = 1.66 x 10^{-27} kg which, in terms of energy, is equivalent to 931.5 MeV or 1.492 x 10^{-10} J (from $E = mc^2$).

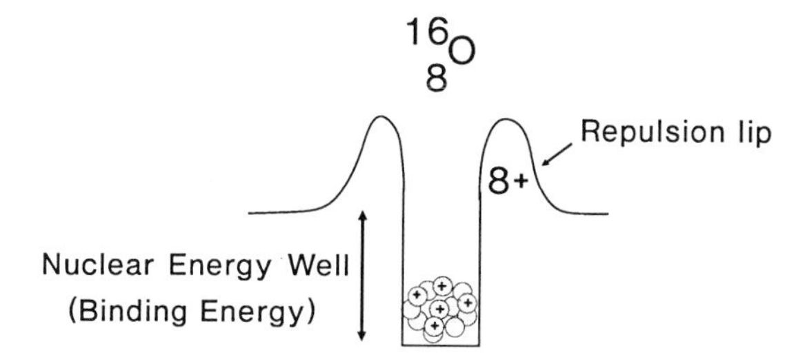

FIGURE 1.2: *The nuclear energy well and electrostatic repulsion lip.*

referred to as the mass defect and gives rise to the binding energy of the nucleus. This binding energy can be depicted by a nuclear energy well in which the nucleons are sitting (*Figure 1.2*), surrounded by an electrostatic repulsion lip. As long as the nuclear energy attractive forces are greater than the electrostatic repulsion forces, the nucleus is stable. In addition to the properties of mass, energy and electrical charge, the nucleons exhibit spin, parity and anti-particle states which affect their interaction in the nucleus.

The number of protons (Z) plus the number of neutrons (N) of an atom is essentially its atomic mass, and is known as the mass number (A), that is, $Z + N = A$. The nucleus of an atom can be described accurately by writing the mass number as a superscript on the left side of its symbol, that is, ^{A}[chemical symbol]. Thus, examples would be ^{1}H, ^{12}C, ^{14}N and ^{16}O.

Whilst the chemical symbol of an atom is synonymous with its chemical properties, the atomic number of an atom describes the number of electrons which is equal to the number of protons (Z). If the atomic number of an atom must be shown, this can be accommodated as a left hand subscript to the chemical symbol, that is, $^{A}_{Z}$[chemical symbol]. Thus, examples would be $^{1}_{1}H$, $^{12}_{6}C$, $^{14}_{7}N$ and $^{16}_{8}O$. The number of neutrons (N) can then be derived from $N = A–Z$. This leaves the right hand side of the symbol free to describe the chemical state of the atom, for example, $^{16}_{8}O_2$.

An atom with its nucleus so accurately described is known as a nuclide. Different nuclides can have different combinations and these are given below:

No. of protons	No. of neutrons	Mass no.	Name	Element
Same	Different	Different	Isotope	Same
Different	Same	Different	Isotones	Different
Different	Different	Same	Isobars	Different

An element can contain just one stable isotope, for example, ^{27}Al, or several stable isotopes, for example, ^{24}Mg (78.99%), ^{25}Mg (10.00%), ^{26}Mg (11.01%); the values in brackets show the relative abundance of magnesium isotopes in nature. Altogether there are 109 elements including the transuranic elements. Of these, 81 are stable elements which have 271 stable isotopes. Of these stable isotopes, 162 contain an even number of neutrons and protons, 55 contain an even number of protons and odd number of neutrons, 49 contain an odd number of protons and an even number of neutrons, and five contain odd numbers of both protons and neutrons. Thus, evenness in the number of nucleons (Z and/or N) is a desirable property for nuclear stability.

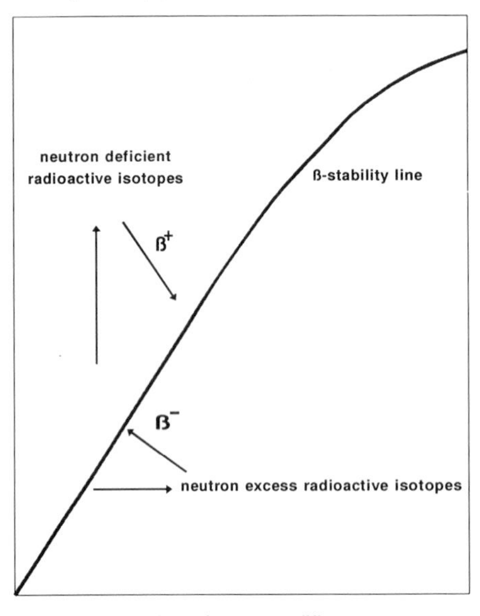

FIGURE 1.3: *'The Chart of the Nuclides': a plot of the number of protons versus the number of neutrons in the nuclides.*

A plot of the number of protons against the number of neutrons in all the stable isotopes reveals another requirement for nuclear stability (see *Figure 1.3*). Up to a mass number of 50, approximately equal numbers of protons and neutrons make for nuclear stability. As the mass number of the nucleus increases above 50, the number of neutrons becomes progressively larger than the number of protons.

In summary, nuclear stability is enhanced by:

(i) A high binding energy per nucleon; a substantial conversion of mass into energy of attraction (binding the nucleons together).

(ii) Even numbers of protons and of neutrons in the nucleus to complete spin pairing.

(iii) An equal number of neutrons and protons in the nucleus, or at least more neutrons than repulsive protons in the nucleus.

Any deviation from these requirements for nuclear stability results in instability of the nucleus, that is, radioactivity.

1.3 Ionizing radiations

The amounts of energy normally associated with physical, chemical and nuclear reactions differ considerably and are shown in *Table 1.2*.

When a large quantity of energy (between a few keV and 8 MeV) is suddenly (approximately 10^{-18} sec) transferred to an atom or molecule, that energy has no time to be dissipated by vibration, rotation and translation between the atoms or molecules. Rather it passes into the atom and onto the electrons. These then jump out of their normal balanced 'positions' and leave the atom. The free electron plus the positive ion left behind are known as an ion pair. This process is known as ionization and takes place along a beam of ionizing radiation, and thus produces an ion track (*Figure 1.4*). When an electron is given a smaller amount of energy, it only jumps or travels a short distance away, before returning to reform the parent atom. This process is known as atomic excitation (*Figure 1.4*).

TABLE 1.2: *Relative energies of different reactions*

Reaction	J/atom	eV/atom	kJ/mol
Kinetic energy of gas molecules	5.3×10^{-21}	0.3	3.2
Chemical changes	4.8×10^{-19}	3.0	289
X-ray generation	8.0×10^{-15}	5×10^{4}	4.8×10^{6}
Nuclear reactions, radioactive decay	1.6×10^{-13}	1×10^{6}	9.6×10^{7}

Ionization and excitation, being extra-nuclear events and associated with electrons, are chemical processes. Thus, the primary absorption reactions in water can be written:

$$H_2O \rightsquigarrow H_2O^+ + e^- \quad \text{ionization,}$$

$$H_2O \rightsquigarrow H_2O^* \quad \text{excitation.}$$

Any particle traveling with sufficient speed (acceleration) constitutes an ionizing radiation. Electromagnetic radiation with a sufficiently low wavelength (high energy quantum) is an ionizing radiation. *Table 1.1* shows some of the ionizing radiations (α^{2+}- and ß-particles, and γ-rays) which come from radioactive isotopes. Their relative properties, and similarity to the fundamental particles of the atom is immediately apparent.

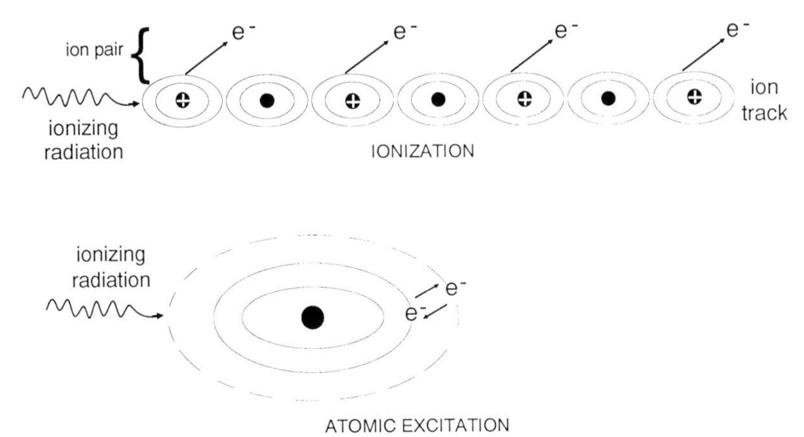

FIGURE 1.4: *Ionization and atomic excitation.*

1.4 Radioactive isotopes (radionuclides)

Radiations capable of ionizing atoms and molecules are known as ionizing radiations. Ionizing radiations can come from outer space, machines or spontaneously out of materials. Materials which emit ionizing radiations are radioactive and such materials can be found in nature, or can be man-made. The molecules and atoms which make up these radioactive materials contain the source of these ionizing radiations. Because heating, cooling and varying the chemical composition (i.e. changing the external electrons) has no effect on the emission of ionizing radiations, they must be coming from deep inside the atom, that is, the nucleus. The source of ionizing radiations are unstable atomic nuclei. An unstable atomic nucleus emits ionizing radiations to gain stability.

Thus, radioactive atoms have unstable nuclei. Those isotopes (nuclides) of an element that emit ionizing radiations are said to be radioactive isotopes (radionuclides). The reasons for this nuclear instability (radioactivity) are the reverse of those given for nuclear stability in Section 1.2 and are:

(i) The nuclear electrostatic repulsion forces become greater than the strong nuclear attractive forces as the number of nucleons in the nucleus increases.

(ii) The number of nucleons in the nucleus is uneven.

(iii) The nucleus contains an excess of, or is deficient in, neutrons.

The latter two causes of radioactivity can be brought about artifically by the introduction of one or more nucleons into a stable nucleus and bringing the weak nuclear forces into play, which result in neutrons changing into protons, and vice versa, depending upon which are in excess.

In nature, all elements with an atomic number (Z) greater than 83 contain only radioactive isotopes. They are naturally radioactive, and are still present today, because they, or their 'parent' radioactive isotope sources, have very long half-lives (>10^8 years). The natural radioactive series begins with ^{232}Th, ^{238}U, and ^{235}Ac, and ends with lead ($Z = 82$). There are naturally occurring long-lived radioactive isotopes with atomic numbers less than 82, but they are few, for example, ^{40}K (0.0119% abundance). Shorter lived ones are formed continuously by cosmic radiation, for example, ^{3}H, ^{10}Be and ^{14}C. In total, approximately 1700 natural and artificial radioactive isotopes are known, which, when plotted with the stable isotopes in *Figure 1.3*, make up the 'Chart of the Nuclides' [1].

A radioactive isotope has three characteristic properties:

(1) The type of ionizing radiations emitted: generally α^{2+}- or $\beta^{\pm}$-particles or γ-rays or a combination.

(2) The rate of radioactive decay measured as a half-life ($t_{1/2}$). The half-life of a radioactive isotope is the time taken for half of the radioactivity to have decayed away. This can take from a fraction of a second to >10^8 years.

(3) The radioactive decay energy: the kinetic energy with which the particles (or the most energetic particles) are emitted, or in the case of γ-rays, the wavelength. Radioactive decay energy is usually expressed in million electron volts (MeV) rather than joules (J). One MeV is equivalent to 1.6×10^{-13} J or 9.64×10^7 kJ/mol.

1.5 Modes of radioactive decay

In the biological sciences the most commonly used radioactive isotopes emit β, or β plus γ, radiation [2]. This is because these radioactive isotopes are the easiest to detect, and present the least radiation hazard. Modes of β and γ decay are therefore discussed prior to α decay.

1.5.1 Beta-decay

All β-decay processes result in a change in the element and very little change in mass. This transformation is due to the 'weak' nuclear forces in the nucleus which permit neutrons to change into protons and vice versa.

(a) In neutron excess radioactive isotopes, this takes the form of β^- (negatron) emission and can be represented by, for example:

$$^{14}_{6}C \rightarrow {}^{14}_{7}N + \beta^- + \tilde{\nu} + \text{energy } (E_{\beta max}),$$

i.e. $$n \rightarrow p^+ + \beta^- + \tilde{\nu} + \text{energy } (E_{\beta max}).$$

neutron proton beta anti-neutrino

The β^--particle and anti-neutrino are formed at the moment of radioactive decay and are thrown out of the nucleus with energy (*Figure 1.5*). The β^--particle has the same properties as an electron traveling at speed and is easily detected. On the other hand, the anti-neutrino carries no charge and virtually no rest mass; its detection is therefore very difficult. More importantly, the anti-neutrino does carry some of the radioactive decay energy so the associated electron carries less energy. β^--particles are therefore emitted with a whole range of energies up to a maximum value ($E_{\beta max}$) when no energy goes with the anti-neutrino (*Figure 1.5*). This maximum value, the radioactive decay energy, is only carried by a few of the β^--particles, the average E_β is approximately one-third of $E_{\beta max}$.

After the nuclear transformation the product nucleus may still be in an excited state and lose that excitation energy by emission of a γ-ray.

(b) In neutron deficient radioactive isotopes, the reverse of the above process takes place, but only if the nuclear instability has sufficient energy available to form two electrons (i.e. > 1.02 MeV). The process is

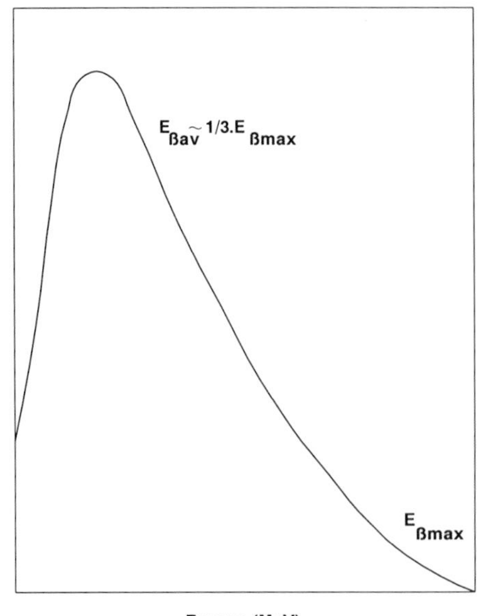

FIGURE 1.5: *A typical β-energy spectrum.*

known as positron emission and can be represented by, for example:

$$^{18}_{9}F \rightarrow ^{18}_{8}O + {}_{1}\beta^{+},$$

i.e. $$p^{+} \rightarrow n + \beta^{+} + \tilde{\nu} + \text{Energy } (E_{\beta max}).$$

proton neutron positron neutrino

As in the case of β^- emission, the positrons (β^+) are emitted with a range of energies up to a maximum of $E_{\beta max}$, the radioactive decay energy. The energy not carried by the positrons is removed by the associated neutrinos. Positrons are electrons with a positive charge, anti-particles which once outside the nucleus and at rest, cannot exist on Earth and are annihilated (see Section 1.8.2).

When there is not enough energy in the nuclear instability to form two electrons, relief from the neutron deficiency is sought by drawing an atomic (extra-nuclear) electron into the nucleus. This process is known as electron capture (EC) and an example is:

$$^{49}_{23}V + {}_{-1}e_{K\text{-}shell} \rightarrow ^{49}_{22}Ti + 4.5 \text{ keV Ti X-rays.}$$

1.5.2 Gamma-decay

Many unstable nuclei are left in an excited state after emission of alpha- (α^{2+}) or beta- (β^- or β^+) particles. This excess nuclear excitation energy is normally lost instantaneously with the emission of an electromagnetic γ-ray. Thus, γ-ray emission is a nuclear de-excitation process and can be depicted by the following example:

$$^{60}_{27}Co \rightarrow \beta^{-} + \tilde{\nu} + E_{\beta max}$$

$$^{60}_{27}Co \rightarrow ^{60}_{28}Ni \rightarrow ^{60}_{28}Ni + \gamma + \gamma$$

radioactive excited de-excited 1.17 MeV 1.33 MeV.

The γ-rays are emitted with definite energies (*Figure 1.6*), confirming the shell structure of the nucleus.

In a very few cases there is an appreciable delay (>l sec) between the emission of the particle and the γ-ray; this is known as internal transition (IT). This is because the nuclear excited state (nuclear isomer) is long lived, that is, it has its own half-life. Radioactive isotopes having this property carry the letter 'm' after their mass number, for example:

$$^{99}_{42}Mo \xrightarrow{t_{1/2}=66\text{ h}} {}^{99m}_{43}Tc + \beta^{-} \xrightarrow{t_{1/2}=6\text{ h}} {}^{99}_{43}Tc + \gamma.$$

The nuclear excitation energy does not always appear external to the radioactive atom which emits it. After emerging from the nucleus, but before leaving the atom, the γ-ray may transfer its energy to an electron. The electron is then emitted from the atom instead of a γ-ray. This process is known as internal conversion (IC). The electrons which arise from it (conversion electrons) are mono-energetic, as opposed to β^--particles which emerge with a range of energies. The empty electron shell in the atom fills up and emits a characteristic X-ray. So instead of a γ-ray being emitted, conversion electrons and X-rays are seen. If these X-rays are in turn internally converted onto the outer atomic electrons, the resultant conversion electrons will have very low (mono-) energy, and are known as Auger electrons.

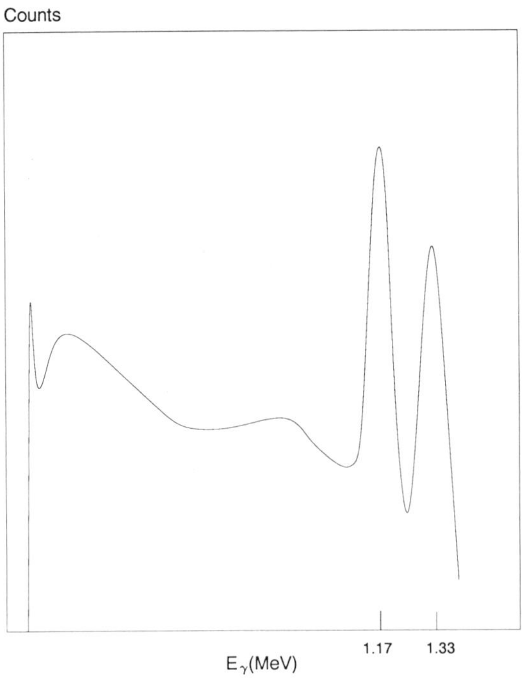

FIGURE 1.6: *Energies of the γ-rays emitted from ^{60}Co.*

1.5.3 Alpha-decay

Alpha-decay arises because there are too many nucleons in the nucleus. The electrostatic repulsion forces are greater than the nuclear attractive forces, so the nucleus wants to lose nucleons, particularly protons. Protons cannot escape alone, and so the emission of the stable α^{2+}-particle is the main form of radioactive decay of the natural radioactive isotopes whose atomic number is >82, that is, above lead in the periodic table. In α-decay, two protons and two neutrons combine in the nucleus to form an α^{2+}-particle, which in itself has great nuclear stability. This can detach itself from the nucleus (rise above the nuclear well) and 'tunnel' its way through the electrostatic repulsion lip which surrounds the nucleus (*Figure 1.2*). The α^{2+}-particle emerges (from the tunnel) with a specific kinetic energy. There is a relationship between the energy of an α^{2+}-emitter and its half-life. This is known as the Geiger–Nuttal rule and shows that the greater the energy with which the α^{2+}-particle is emitted, the shorter is the half-life of the radioactive isotope.

1.5.4 Other decay processes

In the transuranic ($Z > 92$) elements above uranium in the periodic table, the nucleus is very heavy and spontaneous fission (the splitting of the nucleus) becomes a form of decay alternative to α^{2+}-emission. Other unusual singular forms of decay have been reported, for example, delayed emission of neutrons and protons.

1.6 Units of radioactivity

The S.I. unit of radioactivity is the becquerel (Bq), named after Henry Becquerel who discovered radioactivity, and is defined as,

1 Bq = 1 disintegration per second = 1 d.p.s. = 60 d.p.m.,

1 kBq = 1 kilobecquerel = 10^3 d.p.s. = 60 x 10^3 d.p.m.,

1 MBq = 1 megabecquerel = 10^6 d.p.s. = 60 x 10^6 d.p.m.,

1 GBq = 1 gigabecquerel = 10^9 d.p.s. = 60 x 10^9 d.p.m..

Radioactivity is usually expressed as specific radioactivity; in other words, per unit weight or volume, for example, Bq/g, Bq/ml, Bq/mmol, MBq/mol, etc. A non-S.I. unit of radioactivity is the Curie (Ci), named after Marie Curie who discovered radium, which emits radioactivity at the rate of 1 Ci/g = 3.7 x 10^{10} d.p.s./g of radium = 3.7 x 10^4 MBq/g of radium. Becquerels are the preferred units of radioactivity and will be used throughout this book.

When using a counting device to measure radioactivity (see Chapter 3), the results will appear in the form of counts per second (c.p.s.) or counts per minute (c.p.m.). As the counter does not pick up every radiation, c.p.s. and c.p.m. are a fraction of the true d.p.s. and d.p.m. respectively, that is,

$$\text{c.p.s./d.p.s.} = \text{c.p.m./d.p.m.} = f,$$

where f is the fractional efficiency of counting. The efficiency of the counting device must be established with an appropriate calibration source if c.p.s. or c.p.m. are to be converted to Bq. The calibration source should emit radiations of the same type and energy as those to be measured. Counting efficiences are usually expressed as a percentage and thus,

$$\%\ \text{efficiency} = \frac{\text{c.p.s.} \times 100}{\text{d.p.s.}}$$

Knowing the percentage counting efficiency, c.p.s. can be converted to Bq by the equation,

$$\text{Bq} = \text{c.p.s.} \times \frac{100}{\%\ \text{efficiency.}}$$

1.7 Kinetics of radioactive decay

The time taken for an individual unstable nucleus (radioactive isotope) to decay cannot be predicted, nor is it influenced by other nuclei since they are too far apart to affect one another. But looking at a large number (N) of the same unstable nuclei there is a statistical probability that a certain number will have decayed in a given time (t). If N = the number of nuclei, then $-dN$ = the number which will have decayed in time dt. The more nuclei we start with, the more we shall see decay:

$$-dN \propto N;$$

and the longer we observe, the more unstable nuclei will decay:

$$-dN \propto dt.$$

Thus, we can say that,

$$-dN \propto N.dt,$$

which gives rise to the radioactive decay equation,

(1) $$-dN = \lambda.N.dt,$$

where λ is a proportionality constant known as the radioactive decay constant. The number of unstable nuclei (radioactive atoms) decaying in unit time is therefore,

$$-dN/dt = \lambda.N,$$

and thus the amount of radioactivity (A) is,

(2) $$A = -dN/dt = \lambda.N.$$

Equation 1 can be integrated with respect to time to give,

$$-\ln N = \lambda.t - \ln N_0,$$

where N_0 is the number of unstable nuclei (radioactive atoms) at time zero. Rearrangement of this equation gives,

$$\ln (N/N_0) = -\lambda.t.$$

Taking exponentials and rearranging gives,

$$N = N_0.e^{-\lambda t}.$$

Since $A = \lambda.N$ (from *equation 2*), multiplying by λ gives,

(3) $$A = A_0.e^{-\lambda t}$$

(4) $$\text{or, } \ln A = \ln A_0 - \lambda.t.$$

Thus, plotting radioactivity (A) versus time shows an exponential decline of radioactivity with time (*Figure 1.7a*); while plotting $\ln A$ or $\log A$ versus time gives a straight line with a gradient of $-\lambda$ or $-\lambda/2.303$ respectively (*Figure 1.7b*).

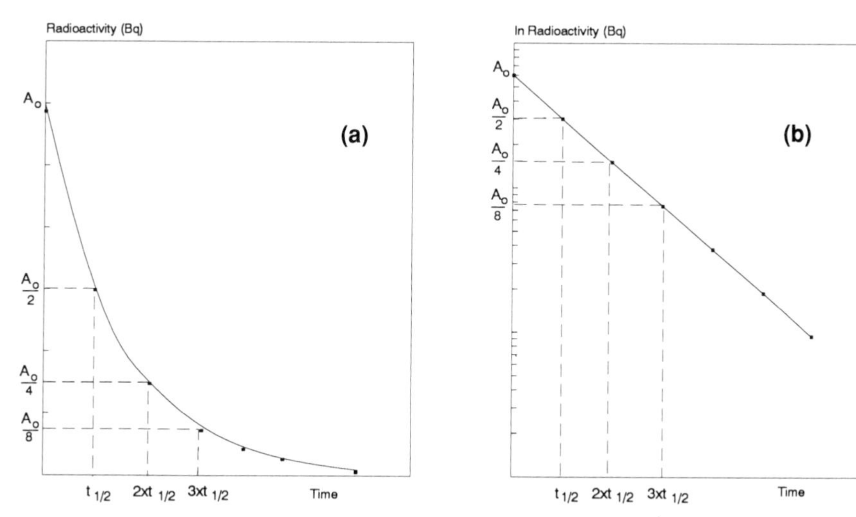

***FIGURE 1.7:** Plot of radioactivity (Bq) versus time on (**a**) a linear scale, and (**b**) a semi-logarithmic scale.*

When half the radioactivity has decayed, $A = A_0/2$, and $t = t_{1/2}$ (the half-life of the radioisotope). Substituting these into *equation 4* gives,

$$\ln \frac{A_0}{A_0/2} = \lambda.t_{1/2}$$

Thus, $$\ln 2 = \lambda.t_{1/2},$$

and, since $\ln 2 = 0.693$, $$\lambda = 0.693/t_{1/2}.$$

Now *equation 3* can be re-written,

(5) $$A = A_0.e^{-0.693.t/t1/2}$$

(6) $$\text{or, } \ln A = \ln A_0 - 0.693.t/t_{1/2}$$

These equations are particularly useful in two repects.

(a) Determination of the half-life of a radioactive isotope. Knowing A_0 and A at any time t after A_0 was measured, the half-life of a radioactive isotope ($t_{1/2}$) can be calculated easily from either equation (5) or (6). Alternatively, from the straight line plot of $\ln A$ versus time, $t_{1/2}$ is the time at which $A = A_0/2$ (*Figure 1.7b*).

(b) Determination of radioactivity in the past and future. Knowing $t_{1/2}$ and A_0 at a particular time, the radioactivity (A) at any time in the future or past can be calculated from equations (5) or (6). Alternatively, from the extension of the straight line drawn between A_0 at zero time and $A_0/2$ at $t_{1/2}$, the radioactivity at any time t can be obtained (*Figure 1.7b*).

If the plot of ln A versus t does not give a straight line, it means that more than one radioactive isotope is present. This is a sign of radioisotopic impurity and the radioactivity measured comes from more than one radioactive isotope. Appendix A shows the half-lives, modes of decay and energies of emitted particles/rays of the principal radioisotopes commonly used in the biological sciences.

1.8 The interaction of ionizing radiation with matter

As described in Section 1.3, ionizing radiations with energy <8 MeV ionize and chemically excite atoms and molecules along discrete tracks. This has considerable implications for the handling of radioisotopes [3]. Initially, ionizing radiations will only affect a minute proportion of the total absorber. Where the incident radiation is particulate (e.g. α^{2+}, β^-, β^+), their energy is lost mainly via inelastic collisions. The loss of energy to form an ion pair from the absorption of ionizing radiation (~ 5.6×10^{-18} J) is much greater than the first ionization potential of most substances (e.g. 2.48 and 2.0×10^{-18} J in N_2 and O_2 respectively). This is because the electron being knocked out by absorption of ionizing radiations is not necessarily from the outermost shell in the atom or molecule. The difference in energy is translated into atomic and molecular excitation.

The interaction of charged particles with the atomic electrical field of absorbing media has been worked out in great detail. The theory takes account of the 'stopping power' of the absorber and the properties of the incident radiation, and gives rise to the linear energy transfer (LET) per unit of track. It has been shown that LET is proportional to:

(i) (charge of the incident particles)^2,

(ii) electron density of the absorber,

(iii) (velocity of incident particle)$^{-1}$.

The ionization density along the track produced by ionizing radiations is the major determinant of the quality factor (QF) used in radiation dosimetry (see Section 1.9).

1.8.1 Interaction of α-particles with matter

Because α^{2+}-particles are so heavy (*Table 1.1*), they push electrons in the absorber out of the way, and travel in short straight ionizing tracks. As α^{2+}-particles emerge from the nucleus with discrete energies, all α-tracks will be of approximately the same length. Ionizations in an α-track are very dense and close together or even overlapping. A 1 Bq source would give 5 MeV/sec and form 1.43×10^5 ions/sec.

In air, all the energy from α^{2+}-particles is lost within a few centimetres. In tissue or water, α^{2+}-particles do not travel beyond a fraction of a millimeter. Thus, α-emitters are not an external hazard as they cannot penetrate the dead part of the skin. However, once absorbed into the body, they are a great internal hazard (QF = 20). Thus, work with large quantities of α-emitters should be restricted to a glove box.

At the end of the α-track, the ionizations per unit track are actually greater than at the beginning. Once the α-particles have slowed down and virtually lost all their energy, they pick up two electrons to become helium:

$$\alpha^{2+} + e^- \rightarrow \alpha^+ \,;\, \alpha^+ + e^- \rightarrow He.$$

1.8.2 Interaction of β-particles with matter

Beta-particles also ionize and excite atoms and molecules of materials through which they pass. However, β-particles are very light in weight when compared to α-particles, they travel much faster, and can lose up to half of their incident energy in a single collision with another electron. Another difference is that β-particles emerge from the nucleus with a range of energies up to the $E_{\beta max}$. This results in the β-particles producing a smaller number of ionizations per unit track. However, they travel much further in the absorbing medium, and over a range of distances, when compared to α-particles of similar energy. In air, an unshielded beta source can be a hazard for up to 5 m depending on the $E_{\beta max}$. In aluminum, which has a density greater than water or tissue, the maximum range is usually less than 1 cm. Thus, β-emitters are both external and internal hazards.

Very energetic β-particles can penetrate near to the nucleus of an absorbing atom and undergo an acceleration–deceleration process, which gives rise to Bremstrahlung X-rays. This form of energy loss is usually less than 5% of that by ionization, and depends on the $E_{\beta max}$ of emission and the atomic number of the absorber. Unlike characteristic X-rays (produced by electron filling of empty inner atomic orbits), Bremstrahlung X-rays cover a whole range of energies and wavelengths. Thus, in addition, β-particles can also produce extra-nuclear electromagnetic radiation.

At the end of a β^- (negatron) track, the β^--particle will add on to the nearest electrophilic atom or molecule. For example, in air, it will give rise to O_2^-, that is,

$$\beta^- + O_2 \rightarrow O_2^-.$$

At the end of a β^+ (positron) track, the β^+-particle is in a hostile environment, as it is an anti-particle, and it is annihilated by the nearest electron:

$$\beta^+(\text{or } e^+) + e^- \rightarrow 2\,\gamma\,(\text{each with } 0.51 \text{ MeV}).$$

Thus, penetrating electromagnetic radiations can arise external to the radioactive emission.

1.8.3 Interaction of γ-rays with matter

Gamma-rays and X-rays are electromagnetic radiations whose wavelength is smaller than the radius of the atom, but larger than the radius of the nucleus. In general, these rays will not be absorbed or interact with absorbers and are extremely penetrating. However, occasionally a cataclysmic event occurs in which the whole of the quantum will transfer its energy to an atomic electron. This electron will then be ejected out of its atomic location, leaving behind a positive ion. Thus, ionization can take place from absorption of electromagnetic radiations. However, because it is an infrequent event, the ionizations occur in discrete volumes far apart in the track of ionization (hundreds of meters apart in air). Approximately 20% of the ejected electrons have sufficient energy to form another ion pair in these volumes.

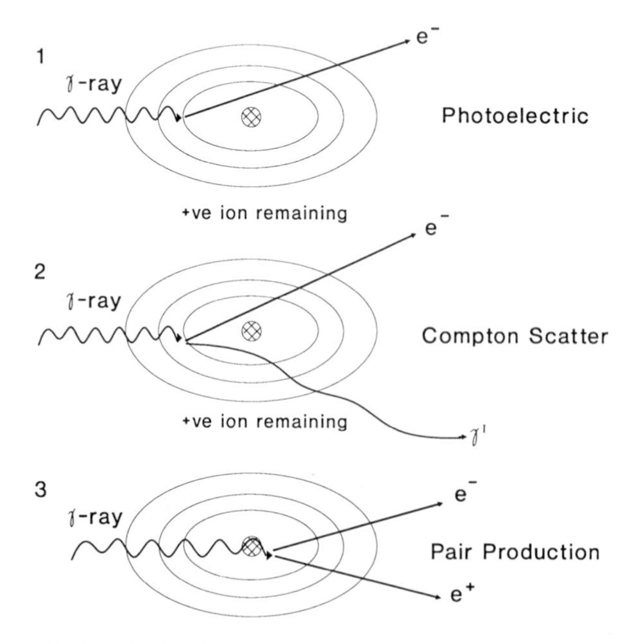

FIGURE 1.8: The interaction of γ-rays with matter.

There are three forms of cataclysmic event depending on the energy of the incident γ-ray; these are photo-electric transfer, Compton scatter and pair production (*Figure 1.8*). The atomic number (Z) of the absorber also affects the form and extent of the interaction processes.

(a) Photoelectric transfer. When a low energy γ-ray strikes an orbital electron, the electron is ejected from the atom with approximately the same energy as the incident γ-ray. The γ-ray is completely absorbed. Ejection of the electron results in ion pair formation.

(b) Compton scatter. This occurs when a γ-ray of intermediate energy strikes an orbital electron which is then ejected from its orbital. Some of the energy of the incident γ-ray is transferred to the electron resulting in ion pair formation, while a secondary γ-ray of lower energy is emitted from the electron with a change of direction.

(c) Pair production. This only occurs when the incident γ-ray is of high energy (>1.02 MeV). In the neighborhood of an atomic nucleus, the reverse of the annihilation reaction occurs and a negatron and positron are formed:

$$\gamma \rightarrow e^- + e^+.$$

The ejected positron is annihilated at the end of its track resulting in the emission of two γ-rays, each of 0.51 MeV.

The conversion of γ-rays to ionizations and atomic excitations by these three processes is best demonstrated by studying the shielding properties of materials against γ-rays. The loss of intensity ($-dI$) of a narrow beam of γ-rays at some fixed distance from the point-source is proportional to the original intensity (I_0) at that point, and also to the thickness (dx) of the absorber placed immediately in front of the γ-source, that is,

$$-dI \propto I_0, \text{ and, } -dI \propto dx.$$

Thus, $$-dI = \mu.I.dx,$$

where μ = the linear absorption coefficient, and is the summation of the photoelectric transfer (τ), Compton scatter (σ) and pair production (k) absorption coefficients (i.e. $\mu = \tau + \sigma + k$), and varies with the incident γ-energy (E_γ).

Thus, $$dI/I = -\mu\, dx.$$

Integration gives, $$\ln I = -\mu.x + \ln I_0,$$

and taking exponentials, $$I = I_0.e^{-\mu.x}.$$

Note the similarity to the decay equations derived in Section 1.7.

1.9 Radiation dosimetry

The absorption of ionizing radiation occurs because the absorber (whether living tissue or inanimate) is present in a radiation field. The source of the radiation may be radioactivity, but it could also be an accelerating machine or natural radiations of the cosmos. Thus, measuring the amount of ionizing radiations absorbed should be independent of radioactivity. Of course, when the strength of a radioactive source is known, it should be possible to establish a link between the two measurements.

The S.I. unit of absorbed radiation is the Gray (Gy). One Gy is defined as one joule per kilogram. A former unit of absorbed radiation was the rad and 1 Gy = 100 rad. Although the absorption of ionizing radiation is measured in a unit relating to the joule, virtually no rise in temperature occurs when radiations are absorbed. As described previously, all the absorbed energy is used up in ionizing and exciting atoms and molecules along the ionizing track. Before radiation (radiolytic) chemical changes can be observed, tens of Gy must be absorbed.

When assessing absorbed radiation in biological systems, the geometric distribution of ions along the ionization track must also be taken into account, that is, the Quality Factor (QF). When the ion pairs are formed close together (as in α-radiations), more biological damage occurs than when they are formed far apart (as in β- or γ-radiation). Thus, 1 Gy of absorbed α-radiation is more harmful to humans than 1 Gy of absorbed β- or γ-radiation. In dealing with the biological effects of ionizing radiations another unit is therefore needed which incorporates these findings, and gives rise to the biological dose-equivalence.

The SI unit of dose-equivalence is the Sievert (Sv) and one Sv is defined as one Gy × QF (A former unit of dose-equivalence was the rem and 1 Sv = 100 rem) The QF is taken as 1 for X- and γ-rays and energetic β-particles, and 20 for α^{2+}-particles. Thus, the same absorbed dose of α-radiation will produce 20 times more biological damage (i.e. is equivalent to 20 times more Sv) than β- or γ-radiation. As α-emitters are seldom used in investigative chemical and biological radioisotope work, the Gy measured can be simply equated to the Sv.

1.10 Radiation protection

Background radiation levels lead to an average dose-equivalence to a person in the UK of approximately 2.5 mSv/annum (see *Table 1.3*). This varies with several factors including location and occupation. Ideally, a member of the public should not absorb a dose-equivalence of more than

1mSv/annum above background which equates to 0.114 μSv/h. According to UK law, a classified radiation worker may not exceed a dose-equivalence of more than 50 mSv in any one year, with a further limitation of 75 mSv in 5 years (see Code of Practice (part 3) of The Ionizing Radiations Regulations, 1985). Assuming a 40 h working week and 50 working weeks in a year, 50 mSv/year is equivalent to 25 μSv/h and 15 mSv/year is equivalent to 7.5 μSv/h.

TABLE 1.3: *Average background radiation absorbed by an individual in the UK*

Sources	Absorbed dose (mSv/annum)
Natural	
Radon and thoron gas from the Earth	1.3
Gamma-rays from solids (e.g. building materials)	0.35
Cosmic rays from outer space	0.3
Radioisotopes in the body	0.3
Artificial	
Medical diagnostic X-rays	0.3
Miscellaneous (e.g. fall-out, occupational, etc.)	≤0.03
	~2.58

Radiation hazards arise externally from natural and artificial β- and γ-emitting radioactive isotopes, and from accelerating machines, and, internally from ingestion and inhalation of all radioactive isotopes. The degree of each hazard depends on the type and amount of radioactivity involved, and the physical and chemical form of the source.

1.10.1 Protection against the external hazard

The three major considerations for protection against external β- and γ-emitting hazards are time of exposure, distance from the source and shielding.

(a) Time of exposure. Clearly, the less time spent in a radiation area, the lower the total accumulated dose-equivalence. Thus, access to radiation areas should be restricted and must be controlled when the dose-equivalence rate is >7.5 μSv/h.

(b) Distance from the source. It is possible to calculate the dose-equivalence rate at a given distance from an unshielded γ-source. This requires a knowledge of the shape of the source, the subsequent flux of the radiations, and whether a narrow radiation beam or a beam plus scattered radiation is falling into the position for which the dose-equivalence rate is to be calculated. Assuming that in a laboratory there are only point sources, and that scatter into the position is negligible, the following formula can be used to estimate the absorbed dose-equivalence rate at a distance of 1 m from a point source of γ-radiation:

$$\text{dose-equivalence rate } (\mu\text{Sv/h}) = 0.14 \times \text{MBq} \times E_\gamma \text{ (MeV)}.$$

Air is a poor absorber of γ-radiations and thus only a geometric factor is operating. The energy absorbed from γ-radiations emitted from a point source varies inversely with the square of the distance from a point source. Thus, at a distance of 0.3 m from the point source:

$$\text{dose-equivalence rate } (\mu\text{Sv/h}) = 1.60 \times \text{MBq} \times E_\gamma \text{ (MeV)}.$$

The dose-equivalence rate from a β-emitter cannot be so precisely calculated. β-radiation is absorbed and scattered in air. This reduces the radiation flux at a given distance in air from a β-source more than from a similar γ-source. At the same time, tissues in the path of β-radiation will absorb more energy than from an equivalent γ-radiation. These contradictory effects result in greater dose-equivalences being absorbed near (30 cm) to an open β-source than a similar γ-source; while the reverse occurs at greater distances. Attempts have been made to find suitable formulae for estimating the dose-equivalence from an open β-source, but these have proved unsatisfactory.

(c) Shielding. It is easy to shield against α- and β-radiations. When inside a tin box, or in solution, β-radiation is no longer an external hazard. However, a precipitate containing an energetic β-emitter in an open filter funnel is an external hazard. Thin metal, glass or plastic shields will offer appreciable protection against MBq quantities of β-radiation.

On the other hand, γ-radiation is difficult to shield against. From a knowledge of the linear absorption coefficient of γ-radiation in a material, the thickness of that material required to reduce the dose-equivalence rate by a factor of 2 or 10 can easily be calculated (see Section 1.8.3 and *Table 1.4*). Interlacing lead bricks are the best form of protection against γ-radiation.

1.10.2 Safety and monitoring of the external hazard

Before beginning work with radioactive isotopes, protection against each hazard must be undertaken. In the laboratory, safety begins with training, wearing laboratory coats and safety spectacles, good smooth washable surfaces, good ventilation, and efficient fume cupboards. Radioactive materials, however small, should never be handled, and should always be moved with tweezers or tongs.

Once work with radioactive isotopes has begun, radiation monitoring of the work area and personnel must be carried out. This is to ensure that the design for safety was correct and that it continues to be so. Radiation

TABLE 1.4: *Thickness of lead and water required to lower the dose-equivalence rate from γ-emitting isotopes*

Energy (MeV)	Thickness (cm) to reduce the dose to: One-half		One-tenth	
	Lead	**Water**	**Lead**	**Water**
0.5	0.4	15.0	1.25	50.0
1.0	1.1	19.0	3.5	62.5
1.5	1.5	20.0	5.0	70.0
2.0	1.9	22.5	6.0	75.0

monitoring of work areas is usually done with portable dose-equivalence rate meters calibrated in the range 1–100 μSv/h. Personnel are monitored with film badges or thermoluminescent dosimeters (see Section 3.5).

1.10.3 Protection against the internal hazard

The internal radiation hazard arises from radioactive isotopes which enter the body. They can do so in three ways: ingestion, inhalation, or through open wounds. Laboratory rules should include 'no drinking, no eating, no smoking, no putting on make-up and no working with open wounds'. Access to the laboratory should be restricted to trained personnel, or personnel undergoing training. Everything should be designed for containment of open, unsealed radioactive sources.

The degree of internal hazard will depend on the radiotoxicity and dose-commitment of the radioactive isotopes absorbed. The relative radiotoxicity can best be obtained from Schedule 2 of the Ionizing Radiations Regulations (1985). This gives the quantities (Bq) of radioactive isotopes above which a laboratory must become a controlled area, and the people working in it become classified workers. These quantities are given with respect to the total amount of radioactivity in the laboratory, the amount on the surfaces and the amount in the air.

To counter the internal hazard, the radioactivity must be contained, the degree of containment depending on the degree of hazard. Where a high degree of hazard is expected (α-emitters or any radioactive dust), work must take place in a negative pressure glove box. A lower degree of containment is to work in an efficient fume cupboard, and the minimum containment is a tray on an open bench in a laboratory with good washable benchtops and flooring. The latter conditions will suffice for most work with MBq quantities of ^{14}C- and ^{3}H-labeled compounds. Trial runs (without the radioactive material) should be carried out to test the containment properties of the apparatus.

The greatest danger when working with open sources occurs when the material is first received and its specific radioactivity is at its highest. It is then that β-emitters, such as ^{32}P-labeled organic phosphates used to radiolabel DNA, or γ-emitters, require both shielding and careful remote handling.

Once the experiments are underway, monitoring is necessary of all surfaces including laboratory coats. This can be done with calibrated count-rate meters or wipe tests. The latter is particularly useful as it indicates whether any contamination is loose on the surfaces (approximately 10% of which can be expected to get into the air). Also, ^{3}H-isotopes are almost impossible to monitor with count-rate meters (because of their low efficiency), and require liquid scintillation counting to detect and measure their presence (see Section 3.5). A piece of polystyrene (which is soluble in scintillation fluid) is the best wipe. Wipe tests are also advantageous when general background radiation levels are high.

If loose radioactivity is present on the surfaces, air monitoring should be carried out. Radioactive gas is not as hazardous as particulate matter which will stick in the alveoli of the lungs. Several cubic meters of air are therefore sucked through a special filter paper which is then analyzed for its radioactive content.

Monitoring personnel exposed to an internal radiation hazard requires the taking of biological samples and their radiochemical analysis. The samples taken depend on the radioactive isotope used and its biochemistry, and can be skin wipes, urine or blood.

1.11 Disposal of radioactive waste

Just as workers using radioactive materials should only absorb a radiation dose 'as low as is reasonably achievable', this requirement is of even greater importance for the general public. Restriction of access to controlled and supervized radiation areas prevents the general public from entering the work place. However, there remains the possibility of the general public being exposed to ionizing radiations from the disposal of radioactive waste materials.

Since radiolabeled compounds are expensive, there is usually no desire to dispose of them in large quantities. Further, radioisotopes used for medical imaging usually have short half-lives. Thus, the disposal of radioactive materials from laboratories and hospitals should not present a long-term hazard. On the other hand, the nuclear energy industry wishes to store extremely large quantities of radioactive waste. Their disposal problems reduce those from laboratories and hospitals to almost insignificance. Nevertheless, the locations of even small amounts of radioactivity must be known if they are not to accumulate into a radiation hazard.

Most countries have laws which deal with the holding and disposal of radioactive waste. In the UK it is the Radioactive Substances Act (1960), which requires the user of radioisotopes to be licensed to hold radioactive material (whether sealed or unsealed), and to be authorized for its disposal in solid, liquid or gaseous form. This Act is enforced by the Pollution Inspectorate of the Ministry of the Environment. It requires the user to display at the work place the license and the authorization to dispose of radioactive waste. A record of all radioactivity kept, decayed, and disposed of, must be kept so that a balanced audit can be carried out at all times.

Sealed solid radioactive sources which are no longer required can usually be returned to the manufacturer. This route of disposal can be discussed at the time of buying the source.

Wherever possible, unsealed radioactive material should be safely stored so that it can decay to disposal levels (ideally $10 \times t_{1/2} = 1/1024$th of its original activity). After decay, low-level solid radioactive waste can be disposed in an ordinary refuse dump at a rate of ~ 1 MBq/month, with up to 400 kBq/month dispersed in 0.1 m^3 and no one item containing more than 40 kBq. Tissues and gloves usually fall into this category. An alternative method is to burn low-level radioactive waste into low-level gaseous waste, allowing dispersal in air without re-entry into buildings. A rate of disposal of <1 MBq/month of ^{3}H and ^{14}C, and <1.5 MBq/month of ^{125}I would be considered reasonable. This form of disposal can be applied to animal carcasses, other potential biohazard materials and large volumes of solvents containing small amounts of radioactivity (e.g. scintillation fluid). Aqueous liquid radioactive waste can be discharged into the main drainage system followed by extensive flushing. The mains drainage system should eventually flow into the sea. It is acceptable to dispose of quite large quantities of radioactivity by this route, for example, 75 GBq/month of ^{3}H and 75 MBq/month of all other radioisotopes.

Special arrangements must be made with the Department of the Environment for the disposal of intermediate-level solid waste (>100 MBq) at specific refuse sites. It may also be possible to transfer the waste and incorporate it with that from another institution which disposes of larger quantities of radioactivity more frequently.

High levels of radioactivity (>100 GBq) should never be used in the biological sciences or discharged into the environment.

References

1. Seelmann–Eggebert, W., Pfennig, G. and Münzel, H. (1974) *Chart of the Nuclides,* 4th Edn. Gesellschaft für Kernforschung mbH, Karlsruhe.

2. Choppin, G.R. and Rydberg, J. (1980) *Nuclear Chemistry: Theory and Applications.* Pergamon Press, Oxford.

3. Keller, C. (1988) *Radiochemistry.* Ellis Horwood, Chichester.

2 Production of Radioisotopes and their Incorporation into Biochemicals

2.1 Production of radioisotopes

Some (natural) radioisotopes can be separated and purified from natural radioactive materials (e.g. Marie Curie's isolation of radium from pitchblend). However, their number, amount and applications in the sciences are limited, and virtually all radioisotopes used nowadays are produced artificially. That means nuclear reactions have to be undertaken to turn stable isotopes into radioactive ones.

The conditions for nuclear stability were established in Section 1.4. It emerged from this that anything done to alter these stable conditions will result in nuclear instability, that is, radioactivity. Thus, making the nucleus of any stable isotope possess excess, or be deficient in, neutrons will result in a radioactive isotope. The subsequent mode of radioactive decay will then be related to the type of instability induced.

2.1.1 Production of neutron-excess radioisotopes

Neutrons do not carry an electrostatic charge and therefore can approach a nucleus without being repulsed. No acceleration of the neutrons is necessary. Indeed, a slow moving neutron can easily slip into a nucleus; the slower its speed, the lower its kinetic energy, and the more likely it is to become incorporated into a nucleus. An analogy is a golf ball approaching the hole at slow speed; it is more likely to drop in than if it was driven at high speed.

The nuclear reactor, when operational, contains large quantities of neutrons. It's fuel rods contain small amounts of fissile materials such as ^{235}U. The heavy nucleus of ^{235}U can capture a neutron and break up into various fission products with the release of an average of 2.5 neutrons/atom plus energy, that is,

$$^{235}U + {}^{1}_{0}n \rightarrow \text{fission products} + \sim 2.5\,{}^{1}_{0}n + \text{energy}.$$

The emitted neutrons can cause fission of other ^{235}U nuclei in the fuel rods. This can be controlled such that for every neutron lost, one neutron is gained. Neutrons emitted into the space between the fuel rods are too energetic and are slowed down by colliding with materials such as carbon (as graphite) or deuterium (as heavy water).

When a target element is introduced into the the core of a nuclear reactor, the slow neutrons will interact with the target nuclei to produce neutron-excess radioisotopes. Three typical types of nuclear reaction are:

(a) (n,γ): neutron in, γ-ray out, no change of element, for example,

$$^{23}_{11}Na + {}^{1}_{0}n \rightarrow {}^{24}_{11}Na + \gamma.$$

(b) (n,p): neutron in, proton out, change of element, for example,

$$^{14}_{7}N + {}^{1}_{0}n \rightarrow {}^{14}_{6}C + {}^{1}_{1}p,$$

$$^{35}_{17}Cl + {}^{1}_{0}n \rightarrow {}^{35}_{16}S + {}^{1}_{1}p.$$

(c) (n, α): neutron in, α-particle out, change of element, for example,

$$^{6}_{3}Li + {}^{1}_{0}n \rightarrow {}^{3}_{1}H + {}^{4}_{2}He.$$

Naturally, the target element for neutron absorption is chosen to give the desired radioisotope as the nuclear reaction product. The chemical form of the target material must be such that the formation of other, unwanted radioisotopes (radioisotopic impurities) is avoided. That is why sodium carbonate or bicarbonate are the target materials for ^{24}Na generation; sodium chloride would also yield ^{35}S from ^{35}Cl by proton emission. As a second best, any other radioisotopes produced should have such a short half-life that by the time the product is used, the radioisotopic impurities will have decayed away.

In a mixture containing both stable and radioactive isotopes of the same element, the term 'carrier' refers to the stable isotopes whilst the radioactive isotope is referred to as the 'tracer'. The 'carrier' is usually present in much larger amounts than the radioactive isotopic 'tracer'. For most applications, 'carrier-free' material is not necessary. In fact, it is undesirable because such small physical quantities of radioactive isotopes (e.g. 1MBq of PO_4^{3-} weighs only 3×10^{-10}g) are easily lost. Obviously, 'carrier-free' material cannot be obtained from (n, γ) reactions, whereas it can be achieved with (n, p) or (n,α) reactions followed by chemical separation and purification.

2.1.2 Production of neutron-deficient radioisotopes

Neutron-deficient radioisotopes are made by the bombardment of a stable target nucleus with charged particles such as protons ($^1_1p^+$), deuterons (d = np^+), or α-particles ($He^{2+} = 2n2p^+$). Because of the electrostatic repulsion lip surrounding the nucleus (see *Figure 1.2*), these particles have to be given tremendous kinetic energies. This can only be achieved with accelerating machines such as Van de Graaff accelerators, cyclotrons and synchro-cyclotrons. Machines of this kind are very expensive and need a permanent staff. Also, the amounts of target material which can be irradiated are small and the half-lives of the resultant radioisotopes are usually very short (e.g. $t_{1/2}$ of ^{18}F = 110 min). It follows, therefore, that neutron-deficient radioisotopes are expensive.

The use of neutron-deficient radioisotopes is hardly commonplace in the laboratory. However, they are used in positron emission tomography (PET), an imaging technique which is attracting increasing interest in nuclear medicine (see Section 3.4.3). An example of the production of a neutron-deficient radioactive isotope is ^{18}F from ^{16}O via a (α,d) reaction,

$$^{16}_{8}0 \; + \; ^{4}_{2}He\,(\alpha) \rightarrow \; ^{18}_{9}F \; + \; ^{2}_{1}np\,(d).$$

2.1.3 Generator produced radioisotopes

Radioisotopes with short half-lives are widely used in nuclear medicine (see Chapter 7). Relatively large quantities (up to 740 MBq) can be given to patients and yet, because of the rapid decay of the radioactivity, the radiation burden is kept to a minimum acceptable level. Such radioisotopes cannot be bought in and have to be generated on site. Radioisotope generators work on the principle of the decay–growth relationship between a long-lived ‘parent’ radioisotope and its short-lived ‘daughter’ radioisotope. This is best illustrated by considering the production of ^{99m}Tc ($t_{1/2}$ = 6 h), a radioisotope commonly used in nuclear medicine, from ^{99}Mo ($t_{1/2}$ = 67 h).

^{99}Mo can be produced from the stable isotope ^{98}Mo by a (n,γ) nuclear reaction,

$$^{98}_{42}Mo \; + \; ^{1}_{0}n \; \rightarrow \; ^{99}_{42}Mo \; + \; \gamma.$$

The majority (85%) of ^{99}Mo decays to ^{99m}Tc by negatron (β^-) emission,

$$^{99}_{42}Mo \; \rightarrow \; ^{99m}_{43}Tc \; + \; _{-1}e\,(\beta^-).$$

Molybdate ions ($^{99}MoO_4^{\,2-}$) bind tightly to alumina and can thus be immobilized on a column of alumina. However, the decay product $^{99m}TcO_4^{\,-}$ (pertechnetate) binds only weakly and can be eluted from the column free of $^{99}MoO_4^{\,2-}$ with physiological saline (see *Figure 2.1*). After elution, the $^{99m}TcO_4^{\,-}$ (daughter) radioactivity starts to grow again in the $^{99}MoO_4^{\,2-}$ column until a new equilibrium is reached, and can be eluted again. Thus, the column can be eluted several times to produce $^{99m}TcO_4^{\,-}$.

Other radioisotopes can be generated in a similar manner using different parent radioisotopes.

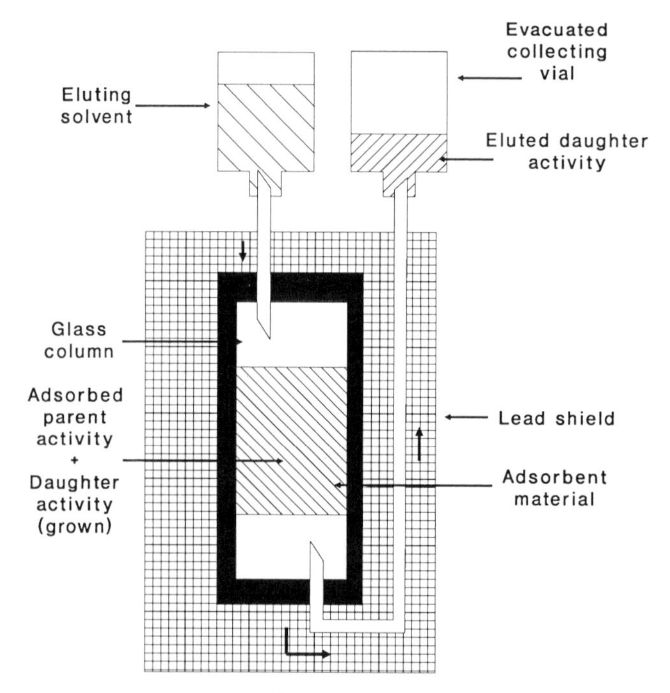

FIGURE 2.1: A typical ^{99m}Tc generator.

2.1.4 The build-up of radioactive material: activation analysis

The amount of radioactive material produced can be calculated because it depends on the difference between the rate of its formation and decay, that is,

$$\text{radioactivity produced} = \text{formation} - \text{decay.}$$

Three factors control the rate of formation:

(i) The number (flux) of the incident neutrons or charged particles.

(ii) The number of target nuclei.

(iii) The 'liking' (resonance) of those nuclei for that particular incident particle and its energy, which is related to the nuclear cross-section.

In activation analysis, the formation of neutron excess or deficient radioisotopes has been adapted into qualitative and quantitative analysis. A non-radioactive sample is made radioactive with a known incident radiation and, from the radioisotopes formed, the target nuclei can be deduced. For example, if a non-radioactive material is irradiated with slow neutrons, and is then found to emit β^--particles of $E_{\beta max} = 0.31$ MeV and two γ-rays of $E_\gamma = 1.17$ and 1.33 MeV, with a $t_{1/2} = 5.26$ years, then the radioisotope must be ^{60}Co (see *Appendix A*). Thus, the non-radioactive sample must have contained ^{59}Co, which was converted to ^{60}Co by a (n,γ) reaction,

$$^{59}_{27}Co + ^{1}_{0}n \rightarrow ^{60}_{27}Co + \gamma.$$

The amount of cobalt per gram of sample can be calculated if all the above factors are known. However, the flux of the incident particles is often difficult to obtain accurately.

Fortunately, a comparative method can be used in which the sample (*X*) is irradiated with the same flux for the same time together with an identical sample to which is added a known quantity of the element (*A*) to be determined. The concentration of *X* in the sample can then be found from:

$$\text{weight of } X \text{ in sample} = \frac{\text{c.p.m.}_{(X)}.(\text{weight of } A)}{(\text{c.p.m.}_{(X+A)} - \text{c.p.m.}_{(X)})}.$$

Activation analysis has been used as a non-destructive method in forensic science to determine, for example, the position and amount of arsenic in human hair.

2.1.5 Radioisotopic purity

Before a radioisotope is used, its radioisotopic purity (also known as radionuclidic purity) should be known. Most radioisotopes are simply detected and measured by the radiations they emit. This measurement may, however, include radiations from other radioisotopes (radioactive impurities) if their radiations are similar. There may also be radiochemical impurities (see Section 2.6), that is, in a different chemical form.

Radioisotopic impurities can arise because:

(a) The target material contains other atoms which can also form radioisotopes.

(b) The prepared radioisotope was not separated and purified from others present.

(c) The radioisotope decays to a radioactive 'daughter'.

All of the above point to the necessity of separating and purifying radioisotopes prior to use.

The three characteristic properties of a radioisotope can be used to test for radioisotopic purity, namely, the type of radiation emitted, the half-life, and the energies with which the radiations are emitted (see Section 1.4). All of these properties can be measured and compared with literature values. Any differences are indicative of a radioisotopic impurity.

As it is very likely that different radioisotopes will be present in the form of a different element or molecule, a test of chemical separation showing radioactivity in more than one fraction will not only indicate a radiochemical impurity, but may also indicate a radioisotopic impurity.

2.1.6 Natural production of radioisotopes

Some biologically important radioisotopes are constantly produced in the Earth's atmosphere by cosmic ray bombardment. For example, when ^{14}N is hit by neutrons, ^{3}H and ^{14}C can be produced by two different nuclear reactions,

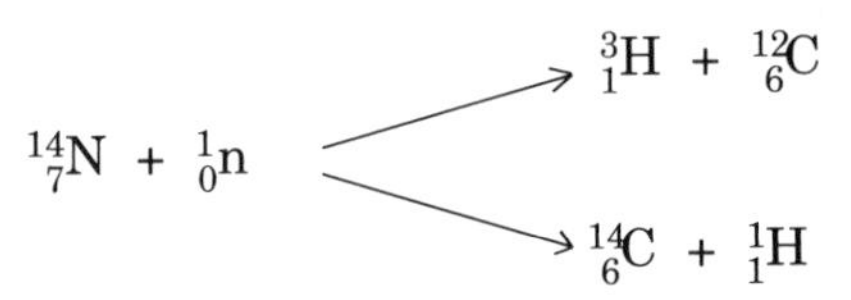

^{3}H and ^{14}C are ultimately incorporated into $^{3}H_2O$ and $^{14}CO_2$ which can enter living cells. Also, ^{40}K (isotopic abundance 0.012%) is an ultra-long-lived radioisotope ($t_{1/2} = 1.26 \times 10^9$ years) and, being water soluble and the major intracellular cation, is present in all cells. These isotopes contribute to the 'background' radiation to which all living systems are exposed (see *Table 1.3*). Indeed, radiocarbon-dating makes use of the decrease in the concentration of ^{14}C by radioactive decay after death and subsequent fossilization. When measuring radioactivity (see Chapter 3), it is important to correct for background radiation, particularly if small amounts are being measured.

2.2 General methods of radiolabeling

In order to radiolabel a biochemical compound, one or more atoms of an element occurring naturally in the compound are replaced by radioactive atoms of the same element. This form of labeling is known as isotopic labeling. The six most common elements in biochemicals are hydrogen, carbon, nitrogen, oxygen, phosphorus and sulfur. There are no practically useful radioisotopes of oxygen and nitrogen. (Although ^{15}N was used in the original demonstration that DNA replication was semi-conservative [1], ^{15}N is a non-radioactive isotope of nitrogen.) Thus, most biochemicals are isotopically labeled with either ^{3}H, ^{14}C, ^{32}P or ^{35}S. Some biochemicals can be made radioactive by 'foreign' or non-isotopic labeling; here, the radioisotope introduced into the molecule does not occur naturally in the compound. The most important biochemical example of 'foreign' labeling is the radioiodination of proteins with ^{125}I and this is discussed in detail in Section 2.3.

It is important for many applications of isotopic labeled compounds that the intramolecular position of the radioisotope is known. This is conventionally indicated in square brackets in front of the chemical name. For example, [1-^{14}C]pyruvic acid describes atoms of ^{14}C present at only C-1 (i.e. -COOH) of the oxo-acid. If the overall labeling is statistically uniform throughout all the atoms of an element in a compound, the labeling is said to be uniform and is represented by the abbreviation U. Thus, [U-^{14}C]pyruvic acid descibes the oxo-acid in which all the carbon atoms have the same statistical chance of being ^{14}C. It does not mean that every carbon atom is ^{14}C. In fact, [U-^{14}C]pyruvic acid could be made by mixing equal radioactive quantities of [1-^{14}C]-, [2-^{14}C]- and [3-^{14}C]pyruvic acid. In such a preparation, only molecules containing a single ^{14}C are present but there is the same statistical chance of the ^{14}C occurring at C-1, C-2 or C-3.

The term isotopic abundance describes the percentage of radioactive atoms present with respect to the total. The theoretical maximum specific activity of ^{14}C at 100% isotopic abundance is 2.31 GBq/mg atom whilst that of ^{3}H is 1.07 x 10^{3} GBq/mg atom. Thus, radiochemicals containing a single ^{14}C or ^{3}H atom per molecule have maximum specific activities of 2.31 and 1.07 x 10^{3} GBq/mmol respectively. Multiple intramolecular labeling with ^{14}C is difficult to achieve and is rarely undertaken. However, if, for example, phenylalanine was prepared by successive radioactive syntheses and every one of the eight carbon atoms in the molecule was ^{14}C, its specific activity would be 18.5 GBq/mmol. In contrast, multiple intramolecular labeling with ^{3}H is much easier and very high specific activities can be achieved. ^{14}C-Labeled steroids are commercially produced with ^{14}C at C-4 only and are supplied therefore at specific activities of up to 2.2 GBq/mmol. ^{3}H-Labeled steroids can be prepared commercially with up to 10 ^{3}H atoms per molecule and are supplied at specific activities up to 5 x 10^{3} GBq/mmol. Such differences in the specific activity of commercially available ^{14}C- and ^{3}H-labeled compounds is of importance when working at nanomolar or picomolar concentrations. If a ^{14}C-labeled compound was diluted to such low concentrations, the amount of radioactivity may fall below detection limits. In such cases, it is necessary to use ^{3}H-labeled compounds which are available at 1000-fold higher specific activities.

Most researchers will buy radiolabeled biochemicals from commercial suppliers. It is, nevertheless, important to appreciate how the radiolabel was introduced into the biochemical. There are two general methods for obtaining radiolabeled biochemicals, namely chemical synthesis and biosynthesis.

Chemical synthesis involves the incorporation of the radiolabel from a chemically simple precursor into a specific position(s) in the molecule. If possible, the radiolabel is introduced during the last stage of the synthesis. Many chemical syntheses give racemic mixtures of optical isomers (e.g. D- and L-amino acids), and great care must be taken to separate and purify only the biochemical isomer. Some examples of radiochemical syntheses are given in *Figure 2.2*. ^{14}C-Labeled steroids are prepared by the reaction between ^{14}C-labeled Grignard reagent (prepared from [^{14}C]methyl iodide) and the product of the oxidative cleavage of the Δ^4,3-oxo system [2]. The resultant steroid is radiolabeled at C-4 only (*Figure 2.2a*). ^{3}H-Labeled steroids are prepared by the reduction of unsaturated precursors with $^{3}H_2$

(*Figure 2.2b*); the number of ^{3}H atoms introduced depends upon the degree of unsaturation of the precursor and the catalyst used [3]. ^{35}S-Labeled L-cysteine can be prepared by reaction of 3-bromo-L-alanine with potassium [^{35}S]thioacetate followed by deacetylation (*Figure 2.2c*).

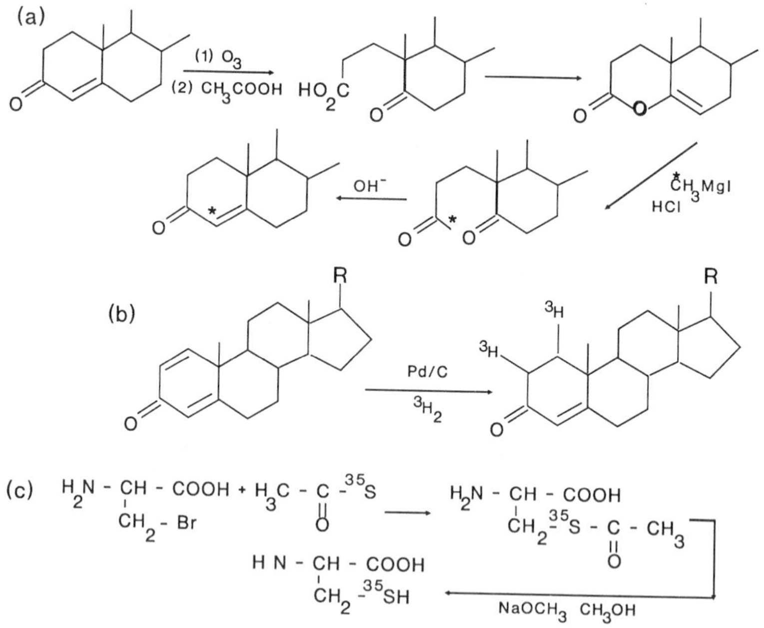

FIGURE 2.2: *Some examples of radiochemical syntheses.(**a**) Preparation of ^{14}C-labeled steroids.(**b**) Preparation of ^{3}H-labeled steroids.(**c**) Preparation of L-[^{35}S]cysteine.*

Biosynthesis uses either micro-organisms or specific enzymes. It is rarely used for ^{3}H-labeling because unforeseen intramolecular migration of ^{3}H atoms is often caused by enzymes (the so-called 'NIH shift'). When micro-organisms are used, the micro-organism is grown in culture with a simple radiolabeled precursor and the desired radiolabeled compound isolated from the organism itself or the cell-free culture medium. Organism biosynthesis produces radiolabeled products of undefined or uniform labeling whereas enzyme biosynthesis produces one, or only a few, products of defined labeling pattern. Photosynthetic micro-organisms such as *chlorella* have been used to produce ^{14}C-labeled sugars when grown with $^{14}CO_2$ as the sole carbon source. D-Glucose produced in this way is uniformly labeled [4]. Similarly, yeast grown on [U-^{14}C]glucose can be used to produce L-[U-^{14}C]amino acids. Several ^{32}P-labeled compounds can be produced by micro-organisms grown on $^{32}PO_4^{3-}$, for example, C_{55}-isoprenyl [^{32}P]phosphate [5].

2.3 Radioiodination of proteins

The incorporation of a radioiodine atom into a peptide, protein or glycoprotein is probably the most common radioisotope labeling procedure undertaken in the laboratory. In almost all cases the iodine atom is 'foreign', that is, it is not part of the structure of the molecule to be radiolabeled. It must be borne in mind that iodination damage can occur, resulting in loss of biological activity of the radiolabeled molecule compared to the unlabeled molecule. This may be caused simply by the very presence of iodine in the molecule, by radiation damage during or after iodination, or by chemical damage caused by the iodinating agents used.

It is theoretically possible to use any of the radioactive isotopes of iodine (^{123}I, ^{125}I, ^{131}I). However, ^{123}I has a very short half-life (13 h) and is of little practical value for laboratory-based work, although it is used in nuclear medicine. ^{125}I is almost always preferred over ^{131}I because of its longer half-life (60 days cf. 8 days), its greater efficiency of counting (80% cf. 30%) and its greater isotopic abundance in commercial sources (90% cf. 20%). In addition, ^{131}I is more hazardous to use than ^{125}I because its γ-rays are more penetrating.

There are two general methods for introducing a radioactive iodine atom into a peptide, protein or glycoprotein, namely direct oxidation or indirect conjugation.

2.3.1 Oxidation methods

These methods rely on introducing ^{125}I into the ortho position of the benzene ring of tyrosine residues in the protein. The oxidation reaction changes the iodide ion from I^- to a positively charged form which is then incorporated into tyrosine residues. Providing relatively small amounts of $^{125}I^-$ are used, monoiodo-tyrosine residues are formed and the resultant product is more stable than if it contained diiodo-tyrosine residues. Ideally, an incorporation of one ^{125}I per protein molecule should be achieved since this will minimize possible tracer damage. A variety of oxidation methods are available and those most commonly used are now discussed.

(a) Chloramine T. The use of chloramine T (the sodium salt of the *N*-monochloro derivative of *p*-toluene sulfonamide) was first described by Hunter and Greenwood in 1962 [6] and is the single most commonly used oxidant for radioiodination. The method is relatively cheap and easy to perform. In aqueous solution, chloramine T produces hypochlorous acid which is thought to oxidize $^{125}I^-$ to the hydrated radical [$^{125}I^+(H_2O)$]. Incorporation of ^{125}I is optimal at pH 7.5 and a strong buffer is necessary to hold the reaction mixture at this pH. The reaction is usually carried out at room temperature and the optimal time of reaction can only be determined by trial and error. Very short reaction times (15 sec) are often used to minimize oxidative damage to the protein. When sufficient ^{125}I has been incorporated, a reducing agent (usually sodium metabisulfite) is added to stop further reaction.

Iodo-beads™ (Pierce Chemical Co.) consist of an insoluble derivative closely related to chloramine T which is coated on to inert beads. These have the advantage that the reaction can be stopped by simply removing the reaction mixture from the beads [7].

(b) Iodo-gen™. This method uses 1,3,4,6-tetrachloro-3α,6α-diphenyl glycoluril as oxidant. This is only slightly soluble in aqueous solution and is therefore less likely to cause oxidative damage to the protein [8]. When coated on to the wall of a plastic tube, the reaction can be terminated by simply removing the reaction mixture. Iodo-gen™ is available commercially either as the free agent or coated on to tubes (Pierce Chemical Co.). An eightfold molar excess of Iodo-gen™ to protein gives maximum incorporation of ^{125}I and the reaction is independent of pH in the pH range from 6.0–8.5.

(c) Lactoperoxidase. Originally reported by Marchalonis [9], this method uses lactoperoxidase and low concentrations of hydrogen peroxide as oxidant. The reaction has to be stopped with a reductant, usually cysteine or mercaptoethanol. More recently, methods have been reported using immobilized lactoperoxidase and where hydrogen peroxide is generated in the reaction mixture by the action of glucose oxidase on glucose. Indeed, such a system is now commercially available (Enzymobead™, Bio Rad Laboratories).

2.3.2 Conjugation methods

In these methods a suitable conjugation reagent containing a phenolic group is first radioiodinated by an oxidative method (usually chloramine T). The ^{125}I-labeled conjugation reagent is then coupled to the protein. Several conjugation reagents have been reported, but Bolton and Hunter reagent (*N*-succinimidyl-3-(4-hydroxyphenyl)propionate) is by far the most commonly used [10]. This can be conjugated (via acylation) to the ε-amino of lysine residues.

^{125}I-HO-C$_6$H$_3$-CH$_2$-CH$_2$-C(=O)-O-N(succinimide) + H_2N-$(CH_2)_4$-CH(NH)(CO) (lysyl residue of protein)

(Iodinated Bolton and Hunter reagent)

→ ^{125}I-HO-C$_6$H$_3$-CH$_2$-CH$_2$-C(=O)-N(H)-$(CH_2)_4$-CH(NH)(CO)

(Iodinated protein)

Alternatively, if the peptide does not contain a lysine residue, conjugation will take place at the free amino group of the N-terminus. Thus, conjugation represents the only method of radioiodinating peptides which do not contain tyrosine residues.

^{125}I-Labeled Bolton and Hunter reagent can either be prepared in the laboratory or purchased directly. The conjugation reaction is carried out under mildly alkaline conditions (pH 8–8.5) and is markedly concentration dependent, so minimum volumes should be used. Bolton and Hunter reagent is unstable in aqueous solution and the reaction is usually performed for 15–30 min at 0°C. Because the reaction does not proceed to completion, unreacted Bolton and Hunter reagent is conjugated to a large excess of a small molecular weight amine (usually glycine).

Irrespective of which labeling method is used, it is necessary to separate the ^{125}I-labeled product from residual $^{125}I^-$ or ^{125}I-labeled Bolton and Hunter reagent (conjugated to glycine). Gel exclusion chromatography is most widely used for soluble proteins and glycoproteins. Ion exchange chromatography can separate $^{125}I^-$ from ^{125}I-labeled proteins and can often separate mono- and multi-iodinated proteins. High performance liquid chromatography (HPLC) provides a better approach for soluble peptides. Indeed, in the case of peptides with more than one available tyrosine residue, HPLC can often separate the reaction products on the basis of which tyrosine residue has been radioiodinated (see Section 4.2.3). If the radioiodinated proteins are associated with cells or subcellular organelles, they can often be simply removed from the iodination mixture by centrifugation and extensive washing.

Several protocols for the radioiodination of peptides, proteins and glycoproteins are described in references [11 and 12].

2.4 Radiolabeling of nucleic acids

Several techniques in molecular biology require the use of radiolabeled nucleic acids, for example, DNA and RNA sequencing, DNA restriction mapping and the detection of specific DNA and RNA sequences by hybridization either *in situ* or on membrane blots. These applications are described in Chapter 6. To label nucleic acids, specific enzymes are used to introduce one or more radiolabeled nucleotides. Since radiolabeled nucleic acids are invariably detected by autoradiography (see Section 3.3), the choice of radioisotope is dictated by the relative needs for resolution and sensitivity. Since ^{32}P is a high energy β-emitter, it gives the greatest sensitivity (but lowest resolution) and is probably the most commonly used isotope for nucleic acid radiolabeling.

Four general methods are available for radiolabeling nucleic acids. Only the principles of these methods will be discussed; detailed protocols are given in references [13 and 14] and kits are available commercially which contain all the necessary reagents for each method. In many cases, it is necessary to remove unincorporated radiolabeled nucleotide from the radiolabeled nucleic acid and this can be achieved by either gel filtration or ethanol precipitation of the nucleic acid.

2.4.1 Nick translation labeling of DNA

Nick translation was the first method described for ^{32}P-labeling of DNA. It depends upon the simultaneous action of two enzymes (see *Figure 2.3*).

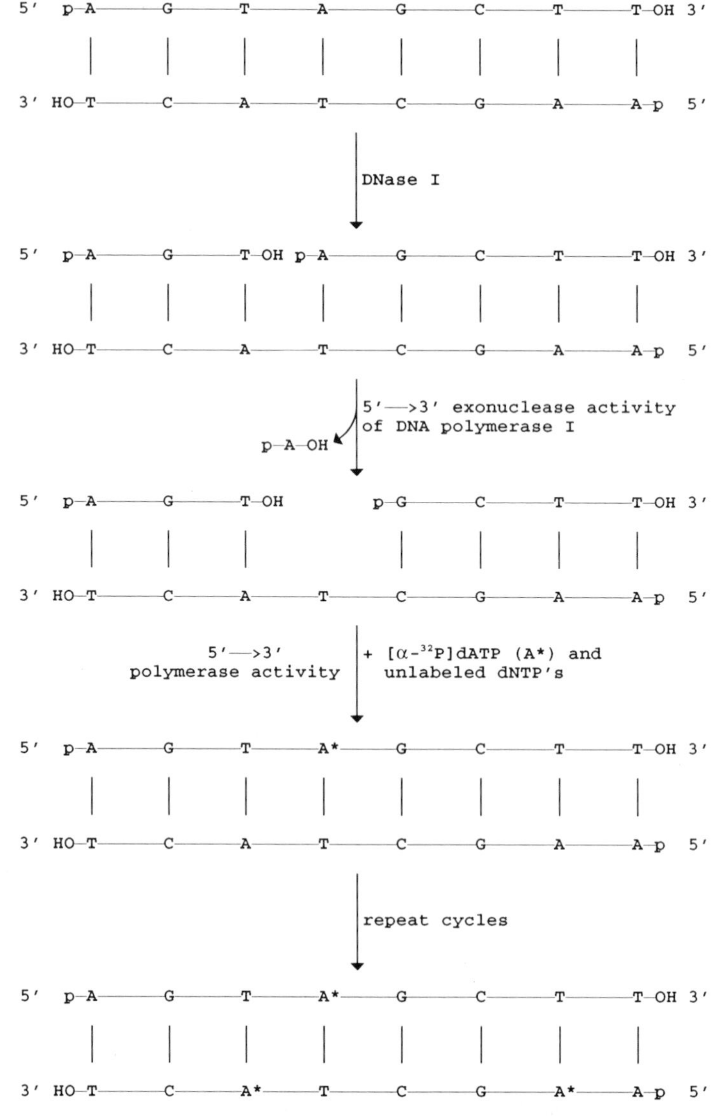

FIGURE 2.3: *Preparation of radiolabeled DNA by nick translation.*

Pancreatic deoxyribonuclease I (DNase I) introduces random nicks in both strands of a DNA molecule producing free 3'-hydroxyl and 5'-phosphate groups at each nick. The 5'–3' exonuclease activity of *E. coli* DNA polymerase I then removes the deoxynucleotides from the 5'-end of the nicks, whilst the 5'–3' polymerase activity adds deoxynucleotides to the 3'-end using the complimentary DNA strand as template. The reaction is carried out at low temperature (approximately 15°C) for 0.5–3 h in an appropriate buffer containing DNase I, DNA polymerase I and all four dNTPs (dATP, dCTP, dGTP and dTTP), at least one of which must be radiolabeled. If a ^{32}P-labeled dNTP is used, it must be radiolabeled in the α-phosphate (i.e. [α-^{32}P]NTP), since this is the phosphate group which is incorporated into the DNA backbone. Similarly, if radiolabeled thiotriphosphates are used, the ^{35}S must be in the α-position. ^{3}H-Labeled dNTPs invariably contain ^{3}H in the purine or pyrimidine base (e.g. [2,8-^{3}H]ATP). Because DNase I introduces nicks randomly, the net result is the formation of a uniform, multiple-labeled population of DNA molecules.

2.4.2 Primer extension labeling of DNA

This method has largely superseded nick translation because it produces DNA of higher specific activity. It relies on the ability of DNA polymerase to synthesize a new strand using single-stranded DNA as template. To initiate synthesis, a free 3'-hydroxyl is provided by a mixture of primers of random base sequence (usually hexamers or nonamers) and results in a uniform, multiple-labeled population of DNA molecules (see *Figure 2.4*).

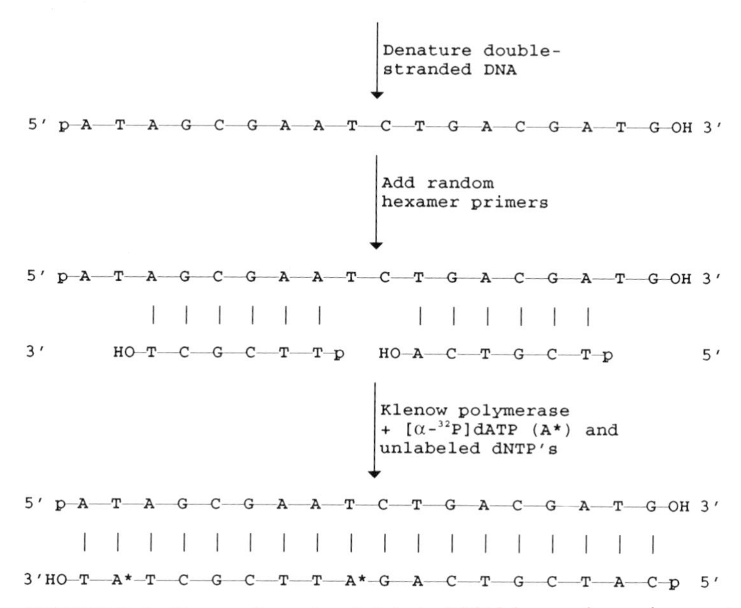

FIGURE 2.4: *Preparation of radiolabeled DNA by random primer extension.*

Alternatively, a unique primer of known sequence can be used to restrict labeling to a sequence of interest. It is essential that the polymerase used is free of 5'–3' exonuclease activity which would otherwise degrade the primer; the Klenow fragment of *E. coli* DNA polymerase I or reverse transcriptase are suitable.

Double-stranded DNA must first be denatured by heating to 95–100°C for 2 min and then rapidly cooled to prevent renaturation. The denatured DNA is then incubated at 20–37°C for up to 3 h with Klenow polymerase and all four dNTPs, one of which must be radiolabeled.

2.4.3 Radiolabeling of RNA

In this method, RNA polymerases are used to synthesize radiolabeled RNA by transcription of a complimentary DNA template in the 5'–3' direction. A vector is required in which a bacteriophage promoter is upstream from a multiple cloning site in the plasmid (see *Figure 2.5*). The most commonly used vectors are pAM 18 and pAM 19 which carry the SP6 and T7 promoters. The double-stranded DNA to be used as template is inserted into the multiple cloning site of the plasmid, which is then linearized by cleavage at a site downstream from the promoter sites. Radiolabeled RNA can now be produced by incubating the linearized DNA template with the relevant RNA polymerase and all four NTPs, one of which must be radiolabeled. [α-^{32}P]UTP is most commonly used and the reaction is allowed to proceed for 1 h at 40°C.

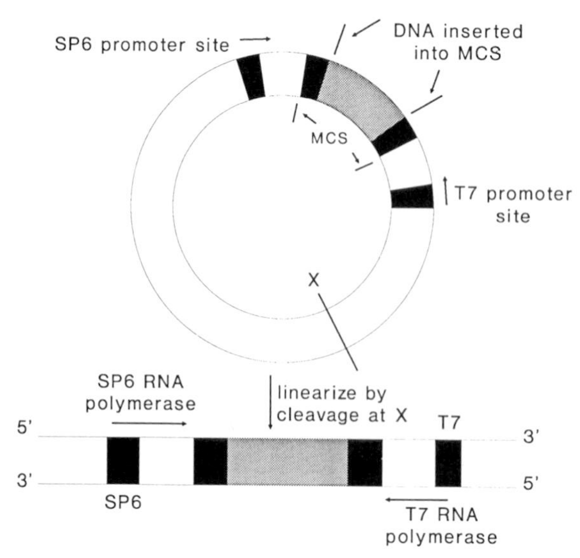

FIGURE 2.5: *Preparation of radiolabeled RNA from DNA inserted into a plasmid containing SP6 and T7 RNA polymerase promoters.*

Because the promoters SP6 and T7 are on opposite sides of the DNA to be cloned, SP6 and T7 RNA polymerase will copy different strands of the linearized DNA template (see *Figure 2.5*). Thus, if both promoters are used, two asymmetric RNA copies are produced. If the resultant radiolabeled RNA is used as a hybridization probe to single-stranded DNA (see Sections 6.4 and 6.5), only one copy will be complimentary and hybridize; this is colloquially known as the 'antisense' probe. The other copy will not hybridize (and is known as the 'sense' probe), and can be used as a control to check for non-specific hybridization.

2.4.4 End labeling of DNA and RNA

All the methods for radiolabeling nucleic acids described so far result in uniform multiple incorporation of the radiolabel. However, some applications of radiolabeled nucleic acids require the incorporation of a single radioactive atom per molecule, for example, the Maxam and Gilbert method of DNA sequencing (see Section 6.1.3). This can be achieved by radiolabeling at either the 3'- or 5'-end. Because such methods incorporate only one radioactive atom per molecule, ^{32}P is always used as it is a high energy β-emitter and confers greatest sensitivity for detection.

3'-End labeling can be achieved by incubating DNA with terminal deoxynucleotidyl transferase (Tdt) and a radiolabeled dideoxynucleoside-5'-triphosphate (usually [α-^{32}P]ddATP). The dideoxynucleotide is added at the 3'-end of the DNA, but, because ddATP lacks a 3'-hydroxyl group, further addition of nucleotides is impossible. Thus, the chain is terminated after elongation by a single radiolabeled base. The reaction is most efficient with single-stranded DNA and requires the presence of a divalent metal cation.

5'-End labeling is achieved by first incubating DNA or RNA with calf intestinal alkaline phosphatase to remove the 5'-phosphate. The nucleic acid is precipitated with ethanol, redissolved in buffer and rephosphorylated by incubation with T4 polynucleotide kinase in the presence of [γ-^{32}P]ATP. Note that, in this case, it is imperative to use ATP labeled with ^{32}P in the γ-position since it is this phosphate group which is transferred to the 5'-hydroxyl group. 5'-End labeling is more popular than 3'-labeling because it can be used on both DNA and RNA, and invariably gives greater incorporation of ^{32}P.

2.5 Technetium (^{99m}Tc) labeling

Because of its short half-life (6 h) and its ease of detection by imaging, ^{99m}Tc is the most common radioactive atom used to prepare radiopharmaceuticals. In order to appreciate how molecules and cells can be radiolabeled with ^{99m}Tc, it is necessary to understand a little of the chemistry of technetium. Technetium has an atomic mass of 99 and is situated in Group VIIa of the periodic table together with manganese and rhenium. It has 5 d-type and 2 s-type valency electrons in its outer orbitals; thus, its highest valency (oxidation) state is +7.

In the laboratory or radiopharmacy, ^{99m}Tc is obtained as the pertechnetate ion ($^{99m}TcO_4^-$) by elution of a ^{99}Mo generator (see Section 2.1.3). The valency state of technetium in $^{99m}TcO_4^-$ is +7, this being the most stable state of the element in aqueous solution. However, in this oxidized state it will not bind with chelating agents or adsorb to particles used for many organ imaging procedures. When ^{99m}Tc is bound to various chelating agents, the reduced states +3, +4 or +5 predominate. Thus, a reduced state must be produced for radiolabeling [15].

In general, the reducing agent used for the production of radiopharmaceuticals is the stannous ion (Sn^{2+}), most often in the form of stannous chloride ($SnCl_2.2H_2O$). In the absence of a coordinating ligand, Sn^{2+} ions in acid solution will reduce Tc(+7) to Tc(+4) according to the following equation,

$$2\,^{99m}TcO_4^- + 16H^+ + 3Sn^{2+} \rightarrow 2\,^{99m}Tc^{4+} + 3Sn^{4+} + 8H_2O.$$

Other reducing agents can be used to produce different valency states, for example, ferric chloride plus ascorbic acid, sodium borohydride, sodium dithionite or concentrated HCl [16].

If the reduction is carried out in the presence of a chelating ligand, that is, a compound that binds to a metal atom by more than one coordinate covalent bond, then ^{99m}Tc complexes will be formed. These form the basis of radiopharmaceuticals used for a range of medical investigations (see Chapter 7).

However, if the proportion of reducing agent to coordinating ligand or the pH of the reaction mixture are incorrect, then the reaction may not go to completion. In such cases, the reaction mixture may contain, in addition to the desired ^{99m}Tc–ligand complex, unreacted (unreduced) $^{99m}TcO_4^-$ ions and the $^{99m}TcO_2^-$ ion (so called hydrolyzed-reduced technetium), which is a colloidal unreactive species. If present, these two impurities will have a different biodistribution within the body to the desired complex.

As a general rule, to limit the appearance of unwanted Tc-species, the reduction of $^{99m}TcO_4^-$ is carried out in acid conditions and under an atmosphere of nitrogen to prevent hydrolysis and oxidation.

2.6 Radiochemical purity

Radiochemical purity is defined as the proportion of the radioisotope present in the stated chemical form and is normally quoted as a percentage. Verification of the purity of a radiochemical may employ several techniques including electrophoresis and chromatography (thin layer, gel exclusion, ion exchange or HPLC) [17]. Successful analyses are dependent on careful technique; reproducibility should be established before credence is given to any particular result.

Having established an initial percentage purity, analyses should be performed at appropriate time intervals to ensure that decomposition will not present a problem if the product is intended to be used over a period of time.

Decomposition of radiochemicals can be brought about by two major causes [18]:

(a) Primary radiation decomposition. As a radioactive atom disintegrates, a fragment of the original molecule will be left behind. If more than one decay process occurs within the same radiochemical, then radioactive fragments are produced, that is, radiochemical impurities [19]. Alternatively, a particle (e.g. a β^+-particle) can interact with the radiochemical and transform it into another species.

(b) Secondary radiation decomposition. The primary effects described above may also have an appreciable effect on other non-radioactive components in the radiochemical preparation. This may result in the production of chemically active species (e.g. free radicals) which will then degrade the radiochemical.

2.7 Radiochemical storage

A first consideration in the storage of radiochemicals is that they must be shielded in a manner appropriate to the type of decay process (see Section 1.10.1). Thus, γ-emitting isotopes such as ^{125}I are routinely stored in lead pots whilst β-emitting isotopes can simply be stored in a tin box or its equivalent.

Practically speaking, little can be done to protect against primary decomposition. Dispersion of the solid compound in a layer thin enough to allow all of the nuclear particles to escape is sometimes useful.

Two methods have been described for retarding secondary radiation decomposition by reducing the formation of free radicals [20]. Firstly, the radiochemical can be dissolved in a solvent (e.g. toluene) which will absorb the primary radiation energy produced by radioactive decay without transferring it to other radiolabeled solute molecules. Secondly, co-solvents can also be used which will react with any free radicals produced in the irradiated solvent (e.g. 2% (w/v) ethanol in water). Thus, many radiochemicals are supplied commercially in solution in either toluene or with a small amount of an alcohol.

Precautions against non-radiation induced decomposition (e.g. physicochemical, photochemical and microbial) are those good housekeeping measures which should be used for storing any biochemical in dilute solution. Thus, it is prudent to store radiochemicals at low temperature (4°C or –20°C) in the dark. Ideally, radiochemicals should be stored in sterile conditions to prevent microbial decomposition, or a bacteriostat could be added.

References

1. Meselson, M. and Stahl, F.W. (1958) *Proc. Natl Acad. Sci.USA,* **44**, 671–682.
2. Milewich, L. and Schweikert, H.U. (1978) *J. Labelled Cmpds Radiopharmaceut.,* **14**, 427–434.
3. Barton, D.H.R., Basu, N.K., Day, N.J., Hesse, R.H., Pechet, M.M. and Starratt, N. (1975) *J. Chem. Soc. (Perkin Trans.),* **1**, 2243–2251.
4. Turner J.C. (1967) *J. Labelled Cmpds,* **3**, 217–233.
5. Stone, K.J. and Strominger J.L. (1972) *Methods Enzymol.,* **28**, 306–309.
6. Hunter, W.M. and Greenwood, F.C. (1962) *Nature,* **194**, 495–496.
7. Markwell, M.A.K. (1982) *Anal. Biochem.,* **125**, 427–432.
8. Fraker, P.J. and Speck, J.C. (1978) *Biochem. Biophys. Res. Commun.,* **80**, 849–857.
9. Marchalonis, J.J. (1969) *Biochem. J.* **113**, 299–305.
10. Bolton, A.E. and Hunter, W.M. (1973) *Biochem. J.,* **133**, 529–538.
11. Bolton, A.E. (1985) *Radioiodination Techniques,* Review 18. Amersham International, UK.
12. Bailey, G.S. (1990) in *Radioisotopes in Biology: A Practical Approach* (R.J. Slater, ed.), IRL Press, Oxford, p. 191–204.
13. Amersham International (1989) *Nucleic Acid Filter Hybridisation.* Amersham International, UK.
14. Cunningham, M.W., Harris, D.W. and Mundy, C.R. (1990) in *Radioisotopes in Biology: A Practical Approach* (R.J. Slater, ed.), IRL Press, Oxford, p. 137–190.
15. Deutsch, E., Libson, K., Jurisson, S. and Lindroy, L.F. (1983) *Prog. Inorg. Chem.,* **30**, 75–139.
16. Nowotnik, D.P. (1990) in *Textbook of Radiopharmacy: Theory and Practice* (C.B. Sampson, ed.). Gordon and Breach, London, p. 53–72.
17. Frier, M., Hardy, J.G., Hesslewood, S.R. and Lawrence, R. (1988) *Hospital Radiopharmacy: Principles and Practice*, IPSM Report No. 56. IPSM, York.
18. Bayly, R.J. and Evans, E.A. (1966) *J. Labelled Cmpds,* **2**, 1–34.
19. Tolbert, B.M., Adams, P.T., Bennett, E.L., Hughes, A.M., Kirk, M.R., Lemmon, R.M., Noller, R.M., Ostwald, R. and Calvin, M. (1953) *J. Am. Chem. Soc.,* **75**, 1867–1868.
20. Bayly, R.J. and Evans, E.A. (1967) *J. Labelled Cmpds,* **3**, 349–379.

3 Measurement of Radioactivity

Most techniques for measuring radioactivity (Bq) can be adapted to detect and measure any of the ionizing radiations, that is, any of the radioactive isotopes. However, some of the radiations react particularly efficiently to give a large signal with certain detectors, and it is to these useful combinations for bio-radioisotope work that attention will be directed (*Table 3.1*).

TABLE 3.1: *Optimum methods of measuring specific radiations*

Type of radiation	Energy	Method
β^--particles (negatrons)	All energies	Liquid scintillation counting Autoradiography
	$E_{\beta max} > 0.7$ MeV	Čerenkov counting
	$E_{\beta max} > 0.2$ MeV	Geiger counting
γ-rays, X-rays and β^+-particles (positrons)	All energies	NaI scintillation counter Autoradiography Gamma camera Ge–Li p–n junction detectors
α^{2+}-particles	All energies	Pulse ionization chamber Zinc sulfide screen

All detection and measurement begins with the ionization and excitation absorption processes in, or on, the surface of the detector. Alpha- and weak β-radiation present a problem because they are easily absorbed, even by air; it follows that emitters of α- and weak β-radiation have to be brought right up to, or better still, right inside the detector. This complication, in addition to the internal health hazard (see Section 1.10.3), is why α-emitters are hardly ever used in the biological sciences.

Unfortunately, weak β-emitters such as ^{14}C ($E_{\beta max}$ = 0.16 MeV) and ^{3}H ($E_{\beta max}$ = 0.018 MeV) cannot be avoided when working with compounds of biological importance. As a result, the most efficient method of measuring such labeled materials is also a very expensive one, namely, liquid scintillation counting.

No matter which method of counting is used, it is important to correct for background radiation (see Section 2.1.6). This is particularly important when measuring low count rates (say <2000 c.p.m.), when background radiation will make a significant contribution to the observed c.p.m..

3.1 Measurement of β^- (negatron) emitters

3.1.1 Liquid scintillation counting

In liquid scintillation counting, the preparation of the sample for measurement is integral to the technique. The material containing the radioisotope is dissolved in the same solvent (usually toluene) containing the scintillator. Chemical excitation is brought about by absorption of ionizing radiation from the β^--emitter dissolved in the common solvent. No matter in which direction the ionizing radiation is emitted, its energy is transferred from solvent molecule to solvent molecule until it is absorbed by a molecule of the scintillator. After excitation by the emitted radiation, the scintillator de-excites to emit, at least in part, quanta of light by fluorescence (i.e. scintillations) (*Figure 3.1*).

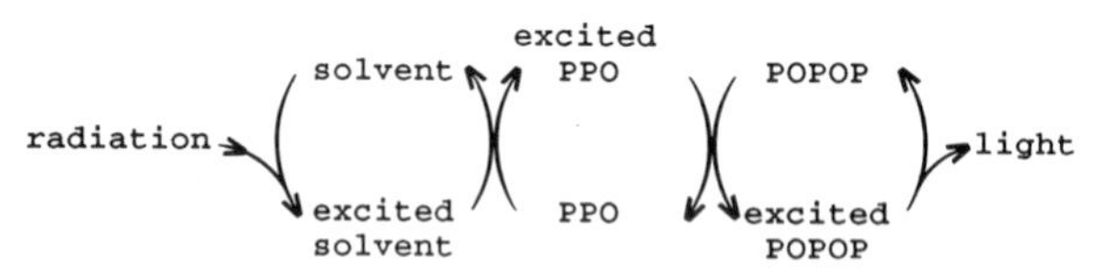

FIGURE 3.1: *From β^--emission to light emission.*

The expense of liquid scintillation counting lies in the computer directed apparatus which automatically passes the liquid sample bottles past a photo-shutter to stand inside a lead shield between two photomultiplier tubes. The photocathodes of the photomultiplier tubes pick up the scintillations which are invisible to the eye and, in less than a microsecond, converts them into electrons and accelerates them along the dynodes to the anode (see *Figure 3.2*). From there, what are now small electronic voltage pulses are amplified (the very small ones are discriminated against) and only allowed to proceed through a coincidence unit if they arrive at the same time. This reduces the background due to spurious pulses beginning on the dynodes of a single photomultiplier tube. Voltage pulses are analyzed by an electronic pulse height analyzer which can be programmed to accept electrical pulses within defined upper and lower limits of pulse height. Different β^--emitters will produce voltage pulses of different pulse heights with a defined scintillator. Thus, the upper and lower limits of the pulse height analyzer can be set to detect specific β^--emitters. Most modern liquid scintillation counters have two, or even three, channels for analyzing samples at different pulse height discriminator settings, such that multiple radioisotopes can be counted at the same time.

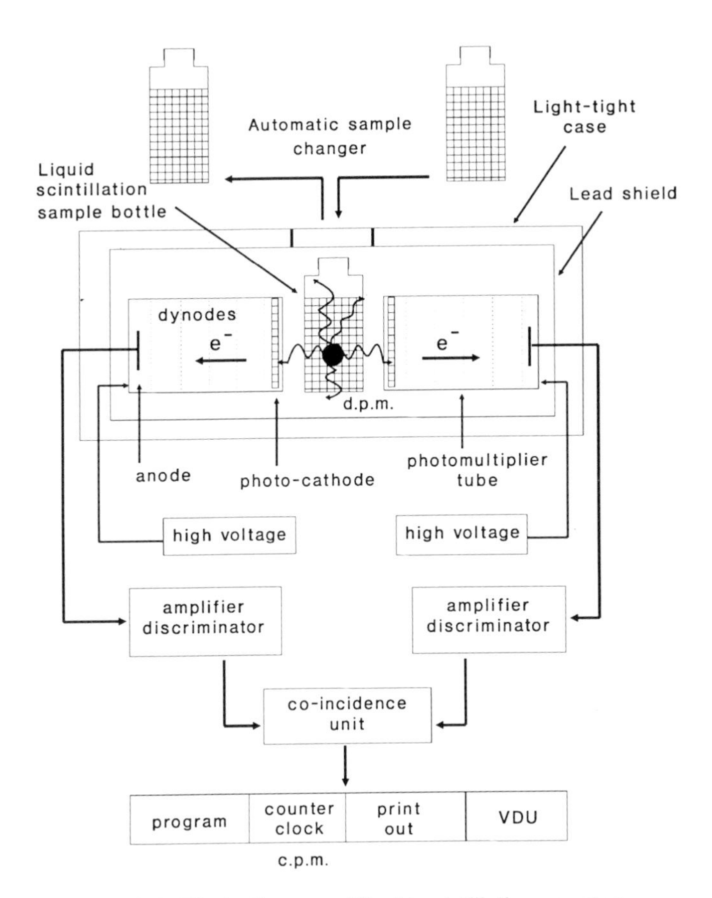

FIGURE 3.2: *Block diagram of liquid scintillation counter.*

Before discussing various aspects of measuring radioactivity by liquid scintillation counting, it is worth mentioning that, since such counters measure quanta of light, any chemiluminesence and bioluminesence in the sample will also be measured.

3.1.2 Sample preparation for liquid scintillation counting

A commonly used primary scintillator is 2,5-diphenyloxazole (PPO) at a concentration of 0.4% (w/v) in toluene [1]. However, the light emitted by PPO is not detected with very high efficiency and 0.02% (w/v) 1,4-di-[2-(5-phenyloxazolyl)]-benzene (POPOP) is routinely added as co-scintillator to improve the efficiency of counting. Thus, an energy transfer process occurs (see *Figure 3.1)*. In this case, a primary and secondary scintillator are required for the simple reason that solvent molecules are incapable of transferring energy directly to the secondary scintillator. Whilst PPO and POPOP were the original scintillators used, and remain a popular choice, other scintillators are now available which do not require a secondary scintillator, for example, 2-(4'-t-butyl-phenyl)-5-(4''-biphenyl)-1,3,4-oxidazole (butyl-PBD).

Many types of scintillation 'cocktail' are available and can be made up in the laboratory or bought from manufacturers. Whilst those based on toluene are the most efficient, they will not accept aqueous liquid samples. Since toluene and water are immiscible, the transfer of energy and/or light is interfered with, resulting in massively reduced counting efficiency. A second solvent mixed with toluene can overcome this problem or alternatively cocktails based on 1,4-dioxane can accommodate aqueous samples up to 10% by volume. A surfactant such as Triton X-100 is often added to scintillation cocktails.

Solid samples such as animal or plant tissues are best counted after solubilization in strongly basic solutions or commercial preparations such as Solvable™ (N.E.N.). Alternatively, tissue samples can be combusted in the presence of oxygen in a commercially available combustion apparatus; ^{14}C-labeled compounds yield $^{14}CO_2$ which is collected in a suitable trapping agent (see Section 5.2), whilst ^{3}H-labeled compounds yield $^{3}H_2O$.

Solid supports such as filter discs, cellulose acetate strips, paper and plastic-backed thin layer chromatograms may be cut out and added directly to a toluene-based scintillation cocktail. This type of counting is known as heterogeneous counting in contrast to homogeneous counting, when the sample is completely dissolved in the cocktail. Counting efficiency is often reduced by self-absorption on to the solid support, particularly for very weak β-emitters such as ^{3}H. However, the convenience of the method far outweighs the reduced efficiency for ^{14}C, ^{32}P and ^{35}S, and even for ^{3}H in many cases.

3.1.3 Liquid scintillation counting efficiency (quenching)

The term quenching refers to any process which reduces counting efficiency. The presence of colored compounds in the sample will reduce counting efficiency if they absorb the quanta of light emitted from the scintillator [2]. Color quenching can often be overcome by bleaching the samples before counting; however, some bleaching agents (e.g. hydrogen peroxide) produce chemiluminesence in some scintillation cocktails. Chemical quenching is due to compounds which interfere with energy transfer from solvent to scintillator, usually as a result of electron capture by the quenching agent.

It is important, therefore, to determine the counting efficiency of each sample, particularly as the efficiency usually varies throughout a series of samples. In this way, observed c.p.m. can be converted to d.p.m. or Bq (see Section 1.6). There are three possible methods for determining counting efficiency:

(a) Internal standardization. This is the original, absolute method for determining counting efficiency, but is little used today. The sample is counted (X c.p.m.) and then a known amount (Z d.p.m.) of the same isotope is added in a small volume of non-quenching solution. The sample is re-counted (Y c.p.m.) and the percentage efficiency calculated from $[(Y - X)/Z] \times 100$. Whilst being relatively simple and reliable, internal standardization is time consuming since it requires accurate and repetitive pipeting of the radioactive standard and each sample has to be counted twice.

(b) Channels ratio. Quenching agents displace the pulse height spectrum to lower energies (see *Figure 3.3*), and this is made use of in determining counting efficiency by the channels ratio method. The pulse height discriminator on one channel of the counter is set to obtain as many c.p.m. as possible from an unquenched standard sample of known d.p.m. (channel A in *Figure 3.3*). The other channel is set to obtain approximately the upper half of the β-spectrum (channel B in *Figure 3.3*). The channels ratio is defined as the c.p.m. in channel B divided by the c.p.m. in channel A, and is always < 1. In practice, this is best carried out in a dual channel counter although, of course, it is possible to count samples twice at different channel settings in a single channel counter.

As increasing amounts of a quenching agent are added to the standard sample, the β-spectrum is displaced to lower energies, increasing the channels ratio (c.p.m. in channel B/c.p.m. in channel A). In addition, the efficiency of counting in channel A can simply be calculated from the observed c.p.m. and known d.p.m. in the standard. Thus, a quench correction curve relating the channels ratio to the counting efficiency can be constructed (*Figure 3.4*). Such curves are specific to the radioisotope and scintillation cocktail used and to the channel settings used. Knowing the channels ratio for a sample, the curve can then be used to determine the efficiency of counting of that sample and hence convert observed c.p.m. to d.p.m. or Bq. In preparing such curves, organic compounds such as aniline, acetone, chloroform, acetic acid, etc., can be added to the standard as quenching agent, or alternatively, standards of known d.p.m. quenched to different degrees can be obtained commercially.

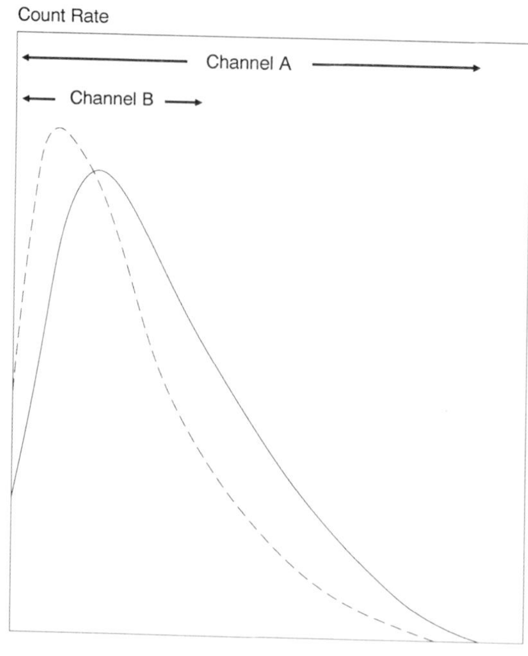

FIGURE 3.3: *The effect of quenching on a β-energy spectrum. The solid line represents unquenched radioisotope and the dotted line represents quenched radioisotope. Channels A and B represent different upper and lower settings on the pulse height discriminator.*

(c) External standardization. This method of determining counting efficiency uses the channels ratio for an external standard and is probably the most commonly used method today. Almost all liquid scintillation counters incorporate a radioactive, γ-emitting external standard (usually ^{226}Ra). After the sample has been counted, the γ-emitting standard is mechanically moved to be adjacent to the sample which is recounted for an additional period. When the γ-rays enter the scintillation cocktail, electrons are produced by pair production and Compton scatter (see Section 1.8.3) and these behave in an identical fashion to the β^--particles emitted in the sample itself. Thus, quenching agents have the same effect on the electrons arising from the absorption of the γ-rays as on the β^--particles in the sample.

The c.p.m. of the external standard is the c.p.m. obtained when the external standard is adjacent to the sample minus the c.p.m. from the sample alone. Knowing the d.p.m. due to the external standard alone, the percentage efficiency with which β^--particles in the sample are counted can easily be calculated. In practice, the external standard is usually counted in two channels at different pulse height discriminator settings to give an external standard channels ratio. Quenched samples will produce higher external standard channels ratios and a correction curve

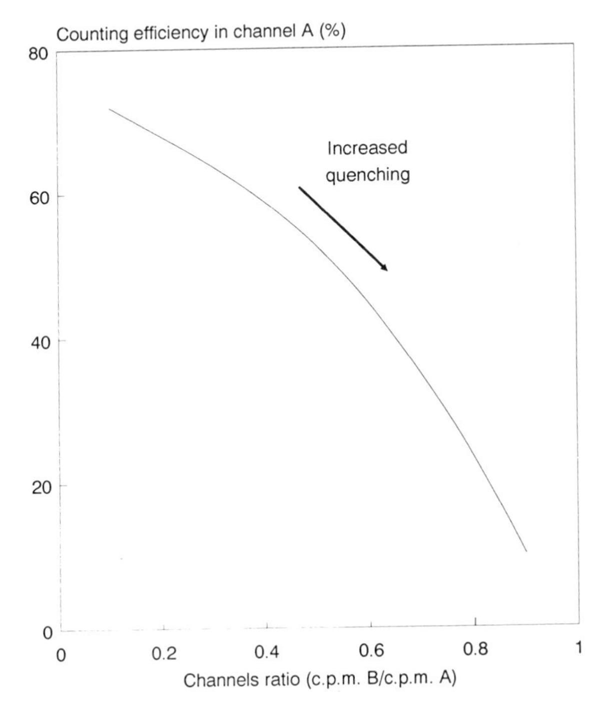

FIGURE 3.4: *A typical channels ratio quench correction curve.*

similar to that in *Figure 3.4* can be produced and used to calculate d.p.m. in the samples. Again, this method of quench correction is specific for the type of scintillation cocktail used and to the channel settings.

3.1.4 Dual label liquid scintillation counting

It is possible to measure simultaneously two different β^--emitting radioisotopes in the same sample, provided their energy spectra are sufficiently different (ideally their average E_β should differ by a factor of at least 10). If the β-spectra of the two radioisotopes only slightly overlap, one channel can be set to count the first radioisotope at an acceptable efficiency whilst rejecting all pulses from the second radioisotope, and the other channel can be set to count the second radioisotope at an acceptable efficiency and reject all pulses from the first radioisotope. However, in practice, this is rarely possible.

A pair of isotopes commonly counted together are ^3H (average $E_\beta = 0.0057$ MeV) and ^{14}C (average $E_\beta = 0.045$ MeV); their energy spectra are shown in *Figure 3.5*. The pulse height discriminator settings on a dual channel scintillation counter are set to obtain a relatively high efficiency of counting for both ^3H and ^{14}C in channel A, and for ^{14}C alone in channel B (see *Figure 3.5*). The counting efficiency of ^{14}C in both channels and of ^3H in channel A must be determined using standards of each separate

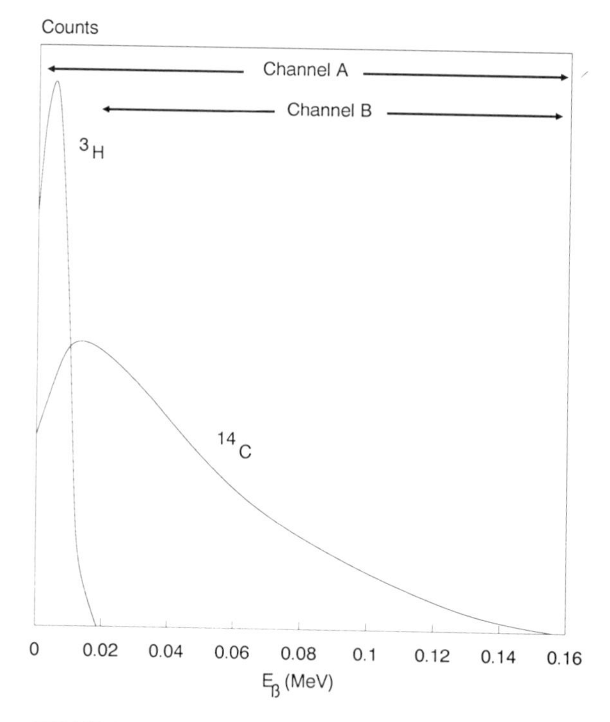

FIGURE 3.5: *β-Energy spectra for ^3H and ^{14}C.*

radioisotope. After counting the unknown dual labeled samples, the d.p.m. of ^3H and ^{14}C can be obtained from the following equations:

$$(\text{c.p.m.})_B = (^{14}\text{C})_{\text{d.p.m.}} \cdot E_{14CB}\,,$$

$$(\text{c.p.m.})_A = (^{3}\text{H})_{\text{d.p.m.}} \cdot E_{3HA} + (^{14}\text{C})_{\text{d.p.m.}} \cdot E_{14CA}\,,$$

where, $(\text{c.p.m.})_A$ = observed c.p.m. in channel A, $(\text{c.p.m.})_B$ = observed c.p.m. in channel B, E_{3HA} = fractional efficiency of counting ^3H in channel A, E_{14CA} = fractional efficiency of counting ^{14}C in channel A, E_{14CB} = fractional efficiency of counting ^{14}C in channel B, $(^3H)_{d.p.m.}$ = d.p.m. of ^3H in sample, $(^{14}C)_{d.p.m.}$ = d.p.m. of ^{14}C in sample.

Other isotopes sometimes counted together are $^3H/^{32}P$, $^3H/^{35}S$, $^{14}C/^{32}P$ and $^{32}P/^{35}S$. Nowadays, many scintillation counters are supplied with preset discriminator settings on each channel for some or all of these common isotope pairs. In addition, the associated computer software will usually calculate the d.p.m. of each isotope.

3.1.5 Čerenkov counting

Radioactive isotopes which emit β^--particles with an $E_{\beta max} > 0.5$ MeV, produce electrons which can travel through water at speeds faster than visible light. Nothing travels faster than light in a vacuum, but in water, glass or transparent plastic, ionizing radiations can travel faster than light if they have sufficient energy. When this happens, part of the slowing down interaction process is the polarization and depolarization of the absorbent molecules along the radiation track, and the accompanying emission of Čerenkov radiation. Čerenkov radiation is a blueish light and can be seen, in the dark, at the bottom of used nuclear fuel storage tanks where the quantities of radioactivity are enormous. The amounts of radioactivity used in laboratory experiments are insufficient for this radiation to be seen with the eye, but the photomultiplier tubes of the liquid scintillation counter can detect and quantify it. Thus, when working with aqueous solutions of a β^--emitter with an $E_{\beta max}$ >0.5 MeV, it is not necessary to make it miscible with, or use, a liquid scintillation cocktail. Instead, the radioactivity in the aqueous solution can be measured directly, after calibration, by its Čerenkov radiation. This method of counting is cheap in that it does not require solvents and scintillators and it does not suffer from quenching problems. *Table 3.2* lists some isotopes suitable for measurement by Čerenkov counting.

TABLE 3.2: *Some isotopes suitable for Čerenkov counting*

Radioisotope	$E_{\beta max}$ (MeV)	% of β-spectrum above 0.5 MeV	Counting efficiency (%)
^{24}Na	1.39	60	30
^{32}P	1.71	80	40
^{36}Cl	0.71	30	10
^{42}K	3.5	90	80

3.1.6 Geiger–Müller counting

By far the cheapest method available for measuring β^--emitters is Geiger–Müller counting. However, to do so efficiently, the isotope must have an $E_{\beta max}$ >0.2 MeV. Thus, ^{3}H ($E_{\beta max}$ = 0.0186 MeV) cannot be measured by this technique whilst ^{14}C ($E_{\beta max}$ = 0.156 MeV) can just about be measured, but only with efficiencies of up to 5%. Geiger–Müller counting depends on the ionization properties of the incident β^--radiation in the gaseous medium of the sealed Geiger–Müller (G–M) tube. *Figure 3.6* is a block diagram of a G–M tube and counter for the measurement of radioactivity in solid samples mounted on aluminum planchettes. These sit in perspex trays on shelves at a defined distance from the G–M tube. G–M counters can also be used to scan thin layer and paper chromatograms. Liquid samples can be dried to produce a thin uniform film. Alternatively, glass G–M tubes with an external annulus or thin-wall dipping G–M tubes can be used.

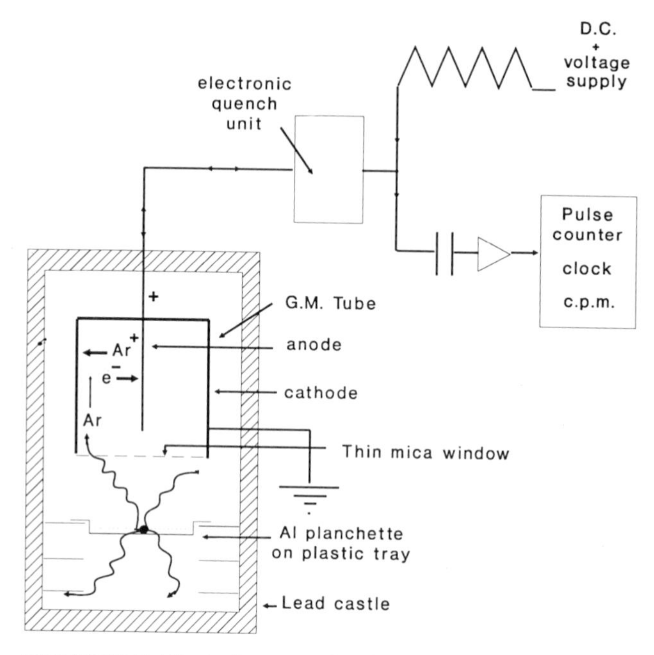

FIGURE 3.6: *Block diagram of a Geiger–Müller counter.*

The G–M tube itself consists of a central anode and a cylindrical cathode with a thin mica end-window through which β^--particles pass. The major gas in a G–M tube is argon and when a β^--particle enters the gas, it displaces an orbital electron from an Ar atom close to its path and forms an ion pair,

$$Ar \rightarrow Ar^+ + e^-.$$

An intense, concentric electrical field is applied to the electrodes which accelerates the electron displaced by the first β^--particle collision to such an extent that it ionizes other gas molecules (gas amplification). These electrons are also accelerated towards the central anode,

$$e^-_{acc} + Ar \rightarrow Ar^+ + 2e^-.$$

Thus, all of the gas around the anode breaks down into a large electric pulse, which can be further amplified and registered over a given period of time, and then expressed as c.p.m.

Because of their relative greater mass, the Ar^+ ions migrate slowly (approximately 200 µsec) to the cathode to pick up an electron and become neutral,

$$Ar^+ + e^- \rightarrow Ar + \text{energy}.$$

However, in doing so, they release more energy which turns into electrons at the cathode, and again begin the acceleration (gas amplification) process towards the anode. Thus, like a neon lamp, once the absorption of radiation switches the G–M tube 'on', it stays 'on'. To avoid this, 10% of the gas in the G–M tube is quench gas (Br_2 or ethanol). The quench gas uses up the energy by decomposing; in the case of bromine, resulting in two free bromine atoms. These eventually recombine, theoretically giving the G–M tube an infinite life-time,

$$Br_2 \longleftrightarrow 2Br\cdot.$$

On the other hand, ethanol-quenched tubes have a life-time limited by the number of ethanol molecules present.

Migration of Ar^+ to the cathode makes for a relatively long (~200 μsec) pulse. The effect is that once a pulse has started to form, further radiations entering the tube cannot produce another voltage pulse during this time, and are therefore not counted. Thus, before the tube can count again it goes through a paralysis time, during which radioactivity is not detected. The paralysis time depends on the type and age of the G–M tube, and can vary from 150–250 μsec. Clearly, the more radioactive the sample, the more pulses are lost and it is important, therefore, that samples for G–M counting are not too radioactive. Knowledge of the paralysis time for a G–M tube allows paralysis correction tables or graphs to be prepared, from which the lost counts can be obtained and added to the observed counts to give the true count rate.

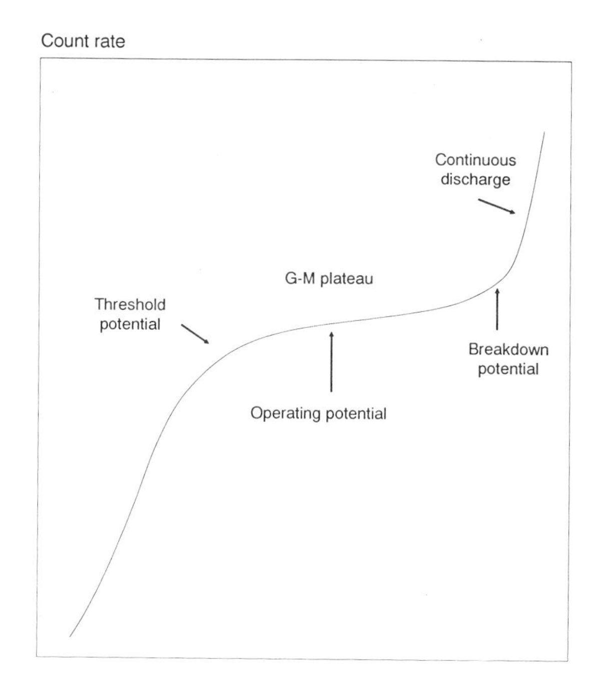

FIGURE 3.7: *Operating voltage characteristics of a Geiger–Müller tube.*

The operating voltage characteristics of a G–M tube are shown in *Figure 3.7*. Count rate initially increases with applied voltage; in this region only the most energetic β^--particles trigger a response. Between the threshold potential and breakdown potential, the observed count rate is approximately constant and it is in this voltage range that G–M tubes are operated (approximately 1000 V for ethanol-quenched tubes or 350 V for Br_2-quenched tubes). At the breakdown potential, the G–M tube continuously discharges and should never be operated in this region otherwise irrepairable damage may occur.

The efficiency of G–M counting is determined by counting a standard sample of the same isotope of known d.p.m. Since the β^--particles originate from a point source, at best only half of the radiation can enter the tube from the planchette. This is further reduced since the probability of a β^--particle entering the tube decreases with increasing distance between the source and G–M tube. Thus, counting efficiency depends not only on the radioisotope being counted, but also on the position of the sample in relation to the G–M tube.

3.2 Measurement of γ-emitters and β^+ (positron) emitters

Gamma-rays (and X-rays) are electromagnetic radiation and are very penetrating. Positrons (β^+-particles) are similar because, as positive antimatter, they are annihilated by negative electrons to form two electromagnetic radiations each with an energy of 0.51 MeV (see Section 1.8.2). Because of the penetrative power of electromagnetic radiations, γ-emitters do not have to be placed inside a detector. In addition, since γ-rays hardly interact with gaseous materials (see Section 1.8.3), solid detectors are more efficient.

The ideal detector for γ-rays is a sodium iodide crystal containing a small amount of thallium as an activator. When γ-rays are absorbed by the transparent sodium iodide crystal, a proportion of the energy absorbed is given out as a flash of light (scintillation). *Figure 3.8* shows a block diagram of a sodium iodide crystal scintillation counter. The sodium iodide crystal is set in a metal 'top hat', lined with a white reflector, and the open end is sealed with perspex. The seal is necessary because sodium iodide is hygroscopic. The perspex faces a photomultiplier tube, which (as in liquid scintillation counting) converts scintillations to a voltage pulse, the height of which is proportional to the energy of the electrons released in the γ-absorption process. The detector and photomultiplier are sealed in a light-tight case, and the source and detector are shielded in lead.

Paralysis correction is not necessary for sodium iodide crystal scintillation counting because scintillations are detected within a microsecond.

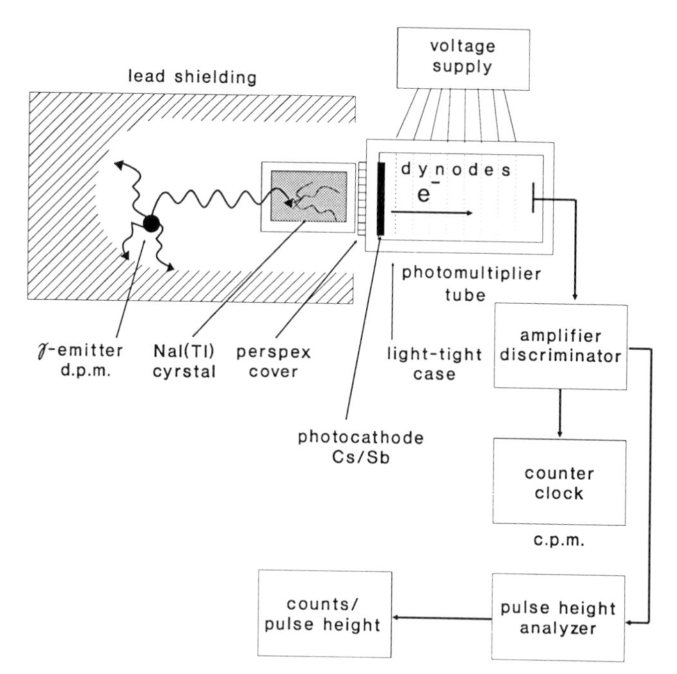

FIGURE 3.8: *Block diagram of a sodium iodide scintillation counter.*

However, the background must still be subtracted, particularly as it is quite high from the sensitive sodium iodide detector (approximately 100 c.p.m.). The efficiency of counting is determined by using standard γ-sources at a given position from the detector. Thus, the unknown radioactivity (in d.p.m. or Bq) of a γ-emitter in a sample can be found.

The sodium iodide crystal scintillation counter can be calibrated to measure E_γ by pulse height analysis. Measuring the number of pulses obtained at different pulse heights from a γ-source results in the formation of a γ-spectrum (see *Figure 3.9*). The higher peaks of the γ-spectrum are due to photoelectric transfer in which the whole of the incident γ-energy (minus a few keV of knock-out energy) is transferred to the ejected electron (see Section 1.8.3). The pulses with lower heights are due to Compton scatter in which only a proportion of the incident γ-energy is passed on to the electron. The electrons released by photoelectric transfer give rise to the brightest scintillations and hence the greatest pulse heights. Thus, by adjusting the pulse height discriminator settings on a sodium iodide crystal scintillation counter, specific radioisotopes can be counted in different channels (in an analogous fashion to liquid scintillation counting). Many counters are available with preset discriminator settings for common isotopes such as ^{51}Cr, ^{60}Co and ^{125}I, or combinations thereof.

Better γ-energy pulse height resolution can be obtained by using germanium–lithium drifted (Ge–Li) p–n junction detectors instead of a sodium

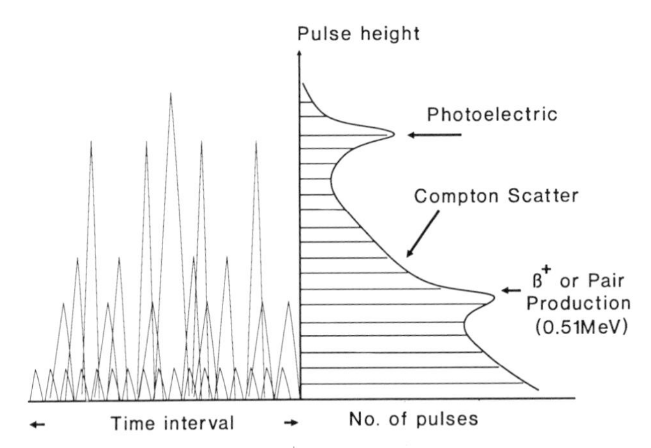

FIGURE 3.9: *Pulse height analysis showing a typical γ-spectrum.*

iodide/thallium crystal. However, in the biological sciences such improved resolution is hardly ever necessary.

3.3 Autoradiography

Radioactive isotopes were originally discovered by their ability to darken photographic film emulsions and, indeed, autoradiography is essentially the detection of radioisotopes using photographic film. In its simplest form, autoradiography consists of placing a flat sample containing radioactivity in close contact with X-ray film in a light-proof container for several hours or even days. Radiation from the areas of radioactivity in the sample forms an image on the film which can subsequently be developed (see *Figure 3.10)*. Autoradiography is used primarily for determining the location of radioisotopes in tissue sections, cells, thin layer chromatograms and polyacrylamide or agarose gels. Nowadays, it is also finding increasing use in the quantitative measurement of radioisotope distribution. A major advantage of autoradiography is that it is non-destructive to the sample.

3.3.1 Direct autoradiography

Photographic emulsions are made of minute grains or crystals of a silver halide (usually silver bromide). When the film is placed in close contact with a flat sample containing radioactivity, β-particles or γ-rays from the sample will hit these crystals reducing many silver ions to metallic silver. When the film is subsequently developed, these silver atoms catalyze the reduction of the entire silver halide crystal to metallic silver to produce an autoradiographic image. This technique, known simply as direct autoradiography, is essentially a single hit process in which all radio-

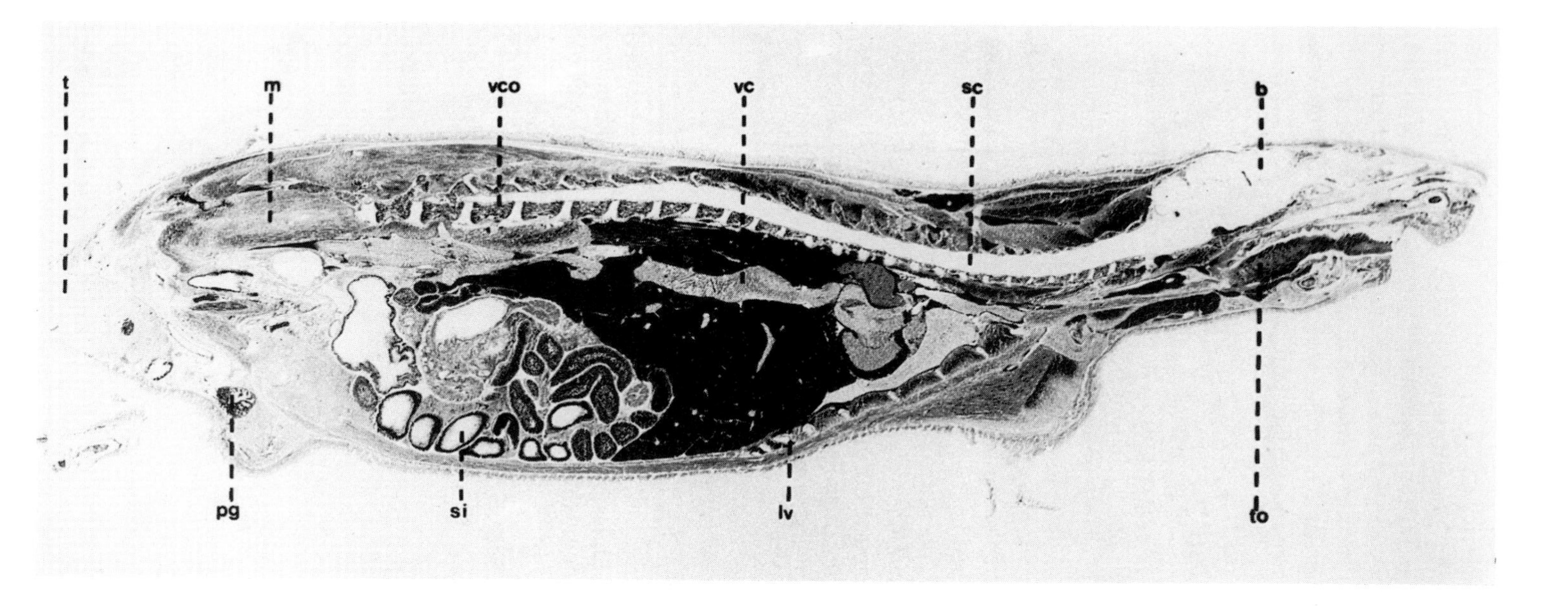

FIGURE 3.10: *A typical whole body autoradiogram showing the biodistribution of a candidate ^{14}C-labeled drug.*

active emissions are recorded equally until all the silver halide crystals are reduced to metallic silver (i.e. the film is saturated). The distribution of radioactivity can be accurately quantified by scanning the developed film in a densitometer. The intensity of the image produced is directly proportional to the amount of radioactivity emitted provided the film is not saturated (which occurs at an absorbance of approximately 1.5). Almost any type of film can be used in direct autoradiography although 'direct' film types such as Kodak Direct Exposure Film™ or Amersham Hyperfilm-β Max™ are most commonly used.

3.3.2 Fluorography

Direct autoradiography gives optimum resolution and moderate sensitivity for β-emitting isotopes with emission energies greater than 0.15 MeV (e.g. ^{14}C, ^{32}P, ^{35}S) and for any γ-emitting isotope (*Table 3.3*). However, direct autoradiography is of little use for weak β-emitting isotopes such as ^{3}H; so little of the emitted energy actually reaches the film that very long exposure times would be required (*Table 3.3*). Similarly, if β-emitters such as ^{14}C or ^{35}S are embedded in a gel matrix, most of the emitted energy is internally absorbed (i.e. quenched) and very long exposure times would again be required. In such cases, sensitivity can be increased without loss of resolution by incorporating an organic scintillator into the sample; this proceedure is known as fluorography. Beta-particles excite the scintillator to produce u.v. light which escapes the sample to produce a photographic image on the film.

TABLE 3.3: *Film detection methods for various isotopes*

Isotope	Type of radiation	Emission energy (MeV)	Range (μm)	Method	Detection limit (d.p.m./cm² for 24 h)
^{3}H	β	0.0186	1	Direct autoradiography	>8 × 10⁶
				Fluorography with PPO	8000
^{14}C/^{35}S	β	0.156/0.167	100	Direct autoradiography	6000
				Fluorography with PPO	400
^{32}P	β	1.71	800	Direct autoradiography	525
				Intensifying screen	50
^{125}I	γ	0.035	>10⁶	Direct autoradiography	1600
				Intensifying screen	100

Data derived from reference [3].

Fluorography is often used to visualize radioactivity in polyacrylamide or agarose gels. A commonly used scintillator is 2,5-diphenyloxazole (PPO). The gel is soaked in PPO dissolved in a suitable solvent followed by soaking in water to precipitate PPO in the gel [4]. Although this method is relatively cheap, it is time consuming. More convenient, but more expensive, commercial preparations are available for impregnating gels with scintillator, for example, Amplify™ (Amersham) and Enlightning™ (Dupont-N.E.N.). In addition, films used in fluorography must have spectral sensitivity optimal for the wavelength of the light emitted by the scintillator. Examples of films suitable for fluorography are Kodak XAR-5™, Amersham Hyperfilm-MP™, Fuji RX™ and Dupont Cronex-4™.

Fluorography carried out at room temperature is not a sensitive technique, and does not produce an image intensity proportional to the radioactivity present in the sample. This is because each photon of light produces only one silver atom when it hits a silver halide crystal. Whilst two or more silver atoms in a silver halide crystal are stable, a single silver atom is unstable and reverts to a silver ion with a half-life of about 1 sec at room temperature. Thus, a latent image capable of subsequent development is only formed when two photons of light hit the same crystal within 1 sec. At low levels of radioactivity, such an occurrence is rare. In order to circumvent these problems, two experimental manipulations are required. Firstly, the sample impregnated with scintillator must be exposed to the film at –70°C; this effectively increases the half-life of silver atoms increasing the time available for a second photon of light to hit the silver halide crystal. Secondly, the film must be hypersensitized before use by exposure to a flash of light (1 msec). This pre-forms a stable pair of silver atoms in each silver halide crystal and each photon of light subsequently hitting the crystal has an equal chance of contributing to the latent image [5]. Thus, exposure to hypersensitized film at –70°C results in an image intensity which is directly proportional to the radioactivity present in the sample.

3.3.3 Intensifying screens

A quite different sensitivity problem occurs with isotopes which emit γ-rays or high energy β-particles such as those from ^{32}P. These emissions pass through and beyond the film so that much of their energy is not recorded on the film (*Table 3.3*). Sensitivity can be increased using an intensifying screen; the film is effectivly sandwiched between the sample and the screen. Intensifying screens contain a dense, inorganic scintillator (usually calcium tungstate) and emit u.v. light in response to irradiation. Thus, a photographic image is effectively superimposed over the autoradiographic image. As with fluorography, exposure should take place at –70°C and use hypersensitized film matched to the wavelength of the emitted light.

Although intensifying screens increase the sensitivity of detection of γ- and some β-emitting isotopes, they do decrease resolution. This is because both the primary emissions (γ-rays or β-particles) and secondary emissions (photons of light) diverge from their source. Thus, by the time secondary emissions hit the silver halide crystals on the film, they may have diverged considerable distances from the source, producing a broad image. Whist this loss of resolution may be unimportant for some samples, for gels containing radiolabeled bands (e.g. DNA sequencing gels) this may become a serious problem.

3.3.4 Non-film methods of autoradiography

Recently, it has become possible to detect and quantify β-emitting isotopes in flat samples without the use of film [6]. An example of this new technology is the Betascope Blot Analyzer which can image patterns of radioactivity from ^{14}C-, ^{32}P- or ^{35}S-labeled gels, blots or thin layer chromatograms. The imaging sensor consists of a gas-filled metal chamber with a thin window in front. Two sets of ionization detection planes are mounted within the chamber (*Figure 3.11*). As β-particles are emitted from the radiolabeled chromatogram mounted on the window, they pass through the chamber leaving a trail of ionized molecules. A short distance into the chamber the β-particles pass through the two measurement planes, each of which consists of three wires maintained at high voltage with respect to one another. The ionization trail sets up a small avalanche of electrons in the vicinity of the wires which is collected on the middle wire. The X/Y coordinate of each ionization event is then recorded from each measurement plane. Since high energy β-particles travel in nearly straight lines, the point of emission from the sample can be calculated by projection from the two detection points back to the source plane. Thus, an autoradiogram-like image can be produced and displayed on a high

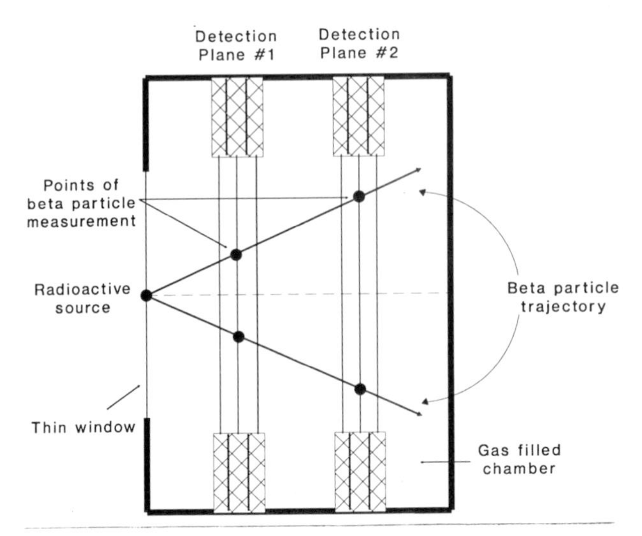

FIGURE 3.11: *The Betascope sensor accurately images radioactivity on the membrane's flat surface by calculating the β-particle trajectory.*

resolution computer monitor. Computer software is available for image quantitation. Such a system has many advantages over conventional film based methods. Images can be produced in as little as 30 min (compared to several hours or days for film) and the amount of radioactivity is directly proportional to the image intensity over all intensity ranges up to ~10^7 d.p.m.. In addition, images can be stored on diskette or tape and recalled at any time. Needless to say, such analyzers are very expensive.

3.4 Medical imaging modalities

During the last 20 years, the use of radioisotopes in medical diagnosis has increased dramatically. This requires the detection of radioisotopes in the body and, in this section, we will discuss the instrumentation available for this.

3.4.1 The gamma camera

The single crystal scintillation camera was first described in 1964 by Anger [7] and is now used to perform most of the high spatial resolution imaging studies performed in nuclear medicine. *Figure 3.12* shows the basic components of a gamma camera. The detector (camera head) produces an image of the distribution of γ-rays emitted from a radiopharmaceutical after its administration to patients. The detector is linked to a console or data processing system which refines the image produced, enabling a hard copy to be portrayed onto photographic film, or stored in a digitized form on an in-built computer. One advantage of using a digitized system is that exposure errors can be avoided and at the same time quantitation can be performed.

The camera head consists of a large diameter (20–50 cm) sodium iodide crystal to which is coupled a matrix of photomultiplier tubes (often 37, or even more in modern gamma cameras). These photomultiplier tubes produce positional signals (X and Y signals) and a Z-pulse, the amplitude of which is governed by the energy of the γ-ray interacting with the crystal.

A collimator (lens) in the face of the crystal permits only radiation coming from a predetermined direction to enter the crystal. The collimator consists of a lead plate through which many small holes have been machined. It can be regarded as a crude filter whereby only those γ-rays traveling in the same plane as the axes of the holes will pass into the crystal. Thus, an image is formed by excluding those γ-rays (the vast majority) which are not traveling in the right direction as they will be absorbed by the collimator. Different collimators are used with variations in hole diameter to govern resolution of the image, and in hole dimensions and number to govern sensitivity.

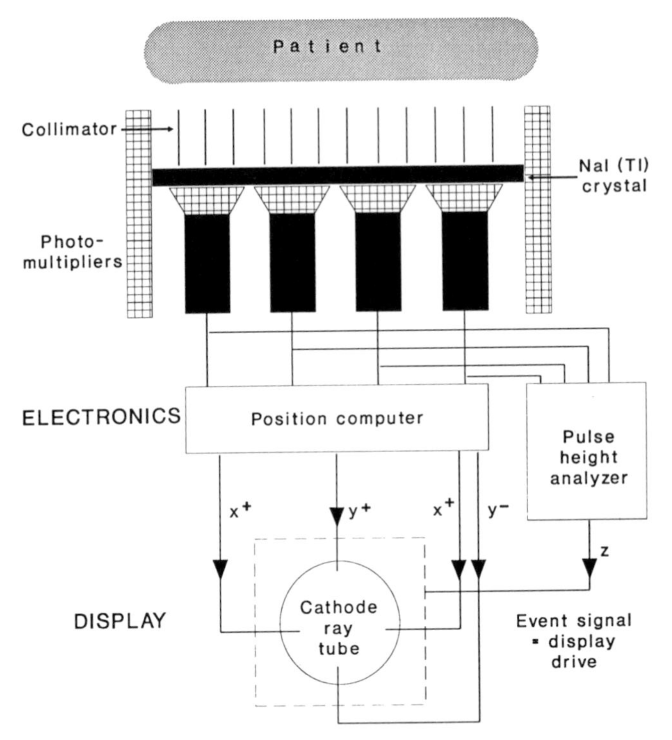

FIGURE 3.12: Block diagram of a gamma camera. Reproduced, with permission of Churchill Livingstone, from reference [8].

The data processing system contains a pulse height analyzer. This allows detection of only those Z-pulses that are representative of the γ-energy of the radioisotope being imaged. By setting the pulse height discriminator around the photopeak of the radioisotope, Z-pulses that are of too high or too low energy (i.e. representing artefactual γ-rays derived from background radiation or Compton scatter) are excluded.

A cathode ray tube and screen display are commonly used as an image display unit. This gives positional and intensity information from the signals produced. Thus, the X and Y signals relate the position on the screen to the position of the incident γ-ray on the face of the crystal, whilst the Z-pulse, if not excluded by the pulse height analyzer, governs the intensity of light on the display. Photographic film may be used to record the intensity of this light.

3.4.2 Single photon emission computed tomography (SPECT)

With the development of imaging computers and their associated mathematics, modern gamma cameras can be used for SPECT imaging as well as conventional (planar) imaging. SPECT enables the portrayal of an image giving positional three-dimensional information (see *Figure 3.13*)

[9]. Planar imaging shows a three-dimensional distribution only as a two-dimensional image and information about depth is lost, and uncertainties will always exist due to tissue overlay.

To perform SPECT, the gamma camera must be able to rotate around the patient. Images are acquired on a step-and-shoot basis (i.e. move to a specific angle, stop, count for a specific time, move to another angle, etc.). Thus, by taking conventional planar views from different directions (typically 64), a three-dimensional image can be built up. Clearly, a computer is essential to store and process all the information necessary to form the three-dimensional image. Conventional SPECT images are viewed in three orthogonal planes; transaxial, sagittal and coronal.

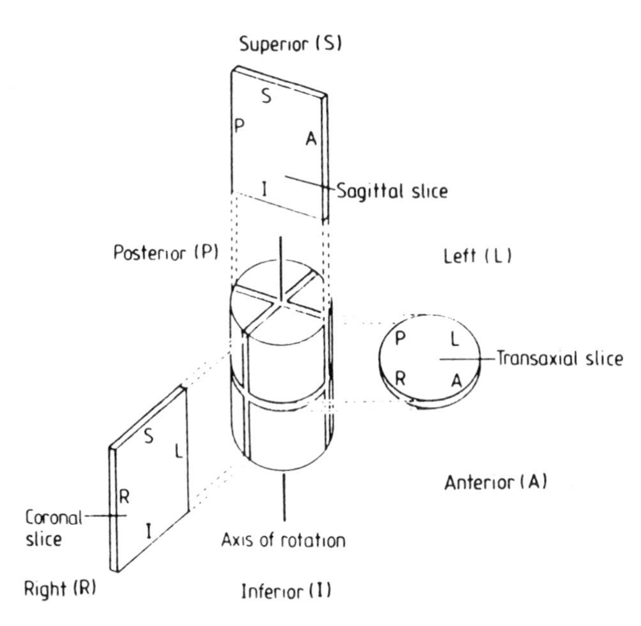

FIGURE 3.13: The orientation of transaxial, sagittal and coronal slices produced in tomographic imaging. By kind permission of Dr R. Ott.

3.4.3 Positron emission tomography (PET)

The three-dimensional image produced by PET is equivalent to that of SPECT in that the reconstructed image is obtained by back-projection of several two-dimensional images taken at different planes. However, unlike SPECT, PET utilizes two rotating γ-ray detectors which do not require collimators [10].

When a positron (β^+-particle) is emitted from a proton-rich nucleus, it travels a short distance before combining with an electron and is then annihilated. This produces two γ-rays, each with an energy of 0.51 MeV, which travel in opposite directions (see Section 1.8.2). When one of the γ-rays interacts with the sodium iodide crystal of one detector, a device known as a coincidence unit is triggered. If a γ-ray is received by the other detector within a specific time interval, the γ-rays are said to be coincident and thus result from a single positron annihilation. By using positional data from the X and Y coordinates of both γ-rays, it is possible to back-project the information along a line joining the detectors, thus giving the position at which the decay process occurs in the patient. The process is then repeated by moving the two detectors through 360° until sufficient information has been acquired to form a three-dimensional image.

The principal advantage of PET imaging over SPECT imaging is that since collimation is not required, the sensitivity of PET is much higher. In addition, the relatively low energy of the incident γ-rays aids reconstruction of the images. However, the major drawback to PET imaging is that the radioisotopes used are all short-lived β^+-emitters and must be produced in a cyclotron. Thus, an on-site cyclotron is necessary for their production.

3.5 Measurement in radiation safety

When working with radioisotopes it is important to assess both the internal and external hazards in terms of Schedules 1 and 2 of the Ionizing Radiations Regulations (1985). Particularly convenient for this purpose are portable, battery-powered count-rate meters with a hand-held probe. These show a very wide (logarithmic) range of counts per second (c.p.s.) directly on a dial. A Geiger–Müller probe with a very thin mica end-window is used for β-emitters such as ^{14}C and ^{35}S. A probe containing a small sodium iodide crystal and photomultiplier is used to monitor ^{125}I and other γ-emitting radioisotopes. By moving the probe of the count-rate meter across a surface (e.g. bench or laboratory coat), large areas can quickly be tested for radioactive contamination. These count-rate meters are based on capacity-resistance circuits which average the counts coming in during a certain time period. Averaging over a short time period (seconds) results in a fast response, but low accuracy. The longest time period which can be spanned is about 6 min, which makes for a slow response, but better accuracy.

Using a γ-emitting point source of known radioactivity and energy, the count-rate meters can be calibrated to read μSv/h (see Section 1.10.1). This effectively converts the instrument into a dose rate meter which can now be used to determine the external hazard; for example, whether a person is standing in a controlled radiation area (>7.5 μSv/h).

Because it is such a weak β-emitter, surface contamination by ^3H cannot be detected by a count-rate meter with a Geiger–Müller probe. Possible ^3H contamination should be monitored by wiping surfaces with either polystyrene or tissue paper moistened in solvent. The wipe is then counted directly in a liquid scintillation counter.

Photographic film badges are a useful long-term monitor of the external hazard and should be worn on laboratory coats of all personnel working with radioisotopes. Film badges from the same batch can be calibrated by exposure to a γ-emitting point source over a long period of time. When film badges are surrendered, they are developed, fixed and scanned with a densitometer to obtain the absorbed personal dose. Thermoluminescent dose meters in the form of finger stalls are also available. Such devices are only required when handling relatively large amounts of strong β-emitters (e.g. ^{32}P) or γ-emitters.

References

1. Fox, B.W. (1976) *Techniques of Sample Preparation for Liquid Scintillation Counting.* Elsevier, North Holland.
2. Peng, C.T., Horrocks, D.L. and Alpen, E.L. (1980) *Liquid Scintillation Counting: Recent Applications and Development. Vol. 1; Physical Aspects. Vol. 2; Sample Preparation and Applications.* Academic Press, New York.
3. Laskey, R.A. (1990) in *Radioisotopes in Biology: A Practical Approach* (R.J. Slater, ed.). IRL Press, Oxford, p. 87–107.
4. Skinner, M.K. and Griswold, M.D. (1983) *Biochem. J.,* **209**, 281–285.
5. Laskey, R.A. (1984) *Radioisotope Detection by Fluorography and Intensifying Screens,* Review 23. Amersham International, UK.
6. Sullivan, D.E., Auron, P.E., Quighley, G.J., Watkins, P.C., Stanchfield, J.E. and Bolon, C. (1987) *Biotechniques,* **5**, 672-678.
7. Anger, H.O. (1964) *J. Nucl. Med.,* **5**, 515–531.
8. Parker, R., Smith P. and Taylor, D. (1978) *Basic Science of Nuclear Medicine.* Churchill Livingstone, Edinburgh.
9. Webb, S., Sutcliffe, J., Burkinshaw, L. and Horsman, A. (1987) *I.E.E.E. Trans. Med. Imaging,* **MI-6**, 67–73.
10. Budinger, T.F., Derenzo, S.E., Gullberg, G.T., Greenburg, W.L. and Huesman, R.H. (1977) *J. Comput. Assist. Tomog.,* **1**, 131–145.

4 Radioimmunoassays and Ligand Binding Assays

Radioimmunoassay and ligand binding assays share many common features. For example, both involve the binding of a radiolabeled molecule to a specific binding site (either an antigen or receptor) and, in both cases, bound radiolabeled molecules must be separated from residual unbound radiolabeled molecules before binding can be quantified. In this chapter we will consider each technique separately although their similarities should become apparent to the reader.

4.1 Radioimmunoassay

4.1.1 Background

The term radioimmunoassay describes the quantitative determination of a biological substance using radiochemical and immunological techniques. The development of radioimmunoassay is attributed to Yallow and Berson [1] who first described the measurement of plasma insulin concentrations by this technique. Nowadays, radioimmunoassay is used to measure concentrations of several compounds including hormones, drugs and their metabolites. The major advantage of radioimmunoassay is that it combines the specificity of immunoassays with the sensitivity of radiochemical methods. As a result, concentrations in the picomolar region can be measured accurately .

Before discussing the principles of radioimmunoassay, it is first necessary to define some immunological terms. An *antigen* is a foreign substance which when administered to an animal will trigger an immune response. The immune response is partly characterized by the synthesis and secretion of an *antibody* from plasma lymphocytes. An antibody is an immunoglobulin protein contained in the γ-globulin fraction of serum proteins which specifically binds to the antigen which caused its production. Serum containing antibodies to a specific antigen is often known as *antiserum*. Only compounds of molecular mass greater than 5 kd are antigenic. Antibodies to compounds too small to be antigenic can be produced by chemically linking the compound (known as a *hapten*) to an antigenic carrier such as bovine serum albumin.

Radioimmunoassay can be used to measure any substance which will serve as an antigen or hapten. It is not an absolute technique and hence requires standards. Thus, the three basic requirements of a radioimmunoassay are pure antigen (as standard), radiolabeled antigen and a specific antibody. The principle of any radioimmunoassay is the competitive binding of radiolabeled antigen and unlabeled antigen to a fixed, limiting number of sites on the antibody. This can be represented:

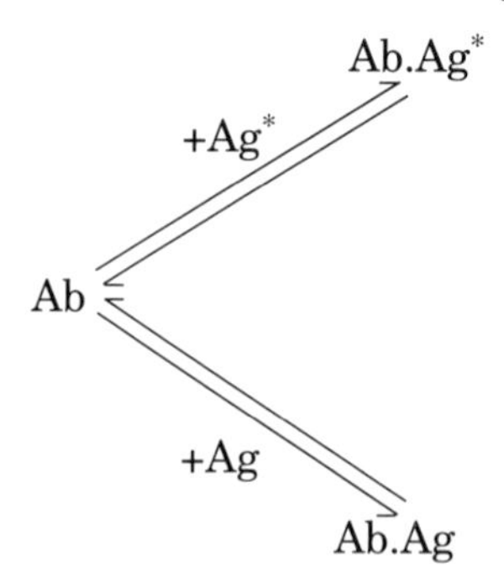

where Ab is the antibody contained in the antiserum, Ag^* is the radioantigen and Ag is the unlabeled antigen present in either standards or unknown samples. At zero concentration of Ag, a large proportion of Ag^* will be bound to Ab. However, the higher the concentration of Ag, the less Ag^* will bind to the fixed amount of Ab because of competition with Ag. By separating the antibody-bound antigen ($Ab–Ag^*$) from the free antigen (Ag^*), the

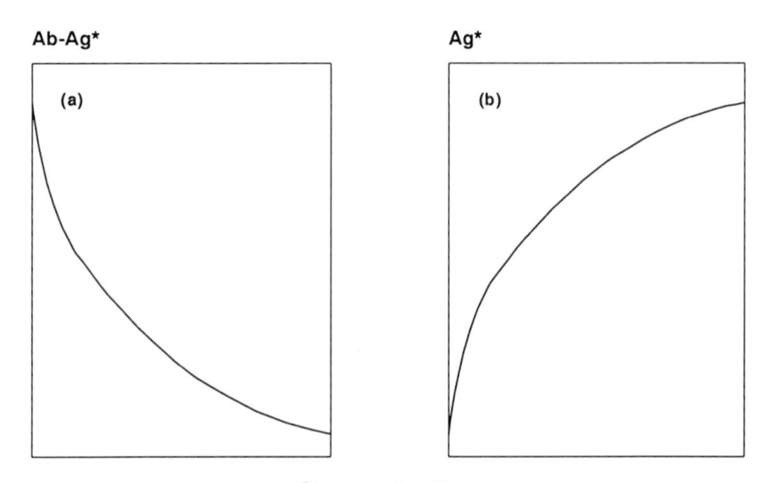

FIGURE 4.1: *Schematic representation of radioimmunoassay calibration curves where,* ***(a)*** *antibody-bound radiolabeled antigen, and,* ***(b)*** *free radiolabeled antigen, are plotted against antigen concentration.*

quantitative determination of the antigen can be made. Counting the radioactivity in the Ab–Ag* complex at various concentrations of Ag will produce a standard curve in which observed radioactivity declines exponentially with increasing antigen concentration (see *Figure 4.1a*). Conversely, counting the radioactivity in Ag* will produce a mirror image standard curve. (see *Figure 4.1b*). Antigen concentrations in unknown samples can then be determined from the appropriate standard curves.

Several steps are involved in the successful development of a radioimmunoassay and these are now discussed in turn.

4.1.2 Production of radiolabeled antigen

The most commonly used isotopes are ^{125}I for protein antigens and ^{3}H for non-protein haptens. Incorporation of these radiolabels into biological compounds has been discussed in Sections 2.2 and 2.3.

Radiolabeled antigens for use in radioimmunoassays do not need to have identical immunoreactivity to the unlabeled antigen being measured. However, although some tracer damage can be accepted, any loss in affinity of the antibody for the labeled antigen will result in some loss of assay sensitivity [2]. Thus, it is preferable to use a labeled antigen that is as near immunologically identical to the unlabeled antigen as possible. This is particularly important for ^{125}I-labeled proteins whose structure is often affected by the iodination process.

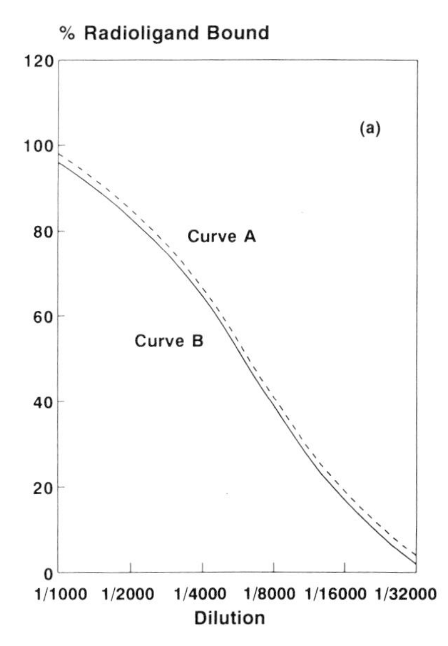

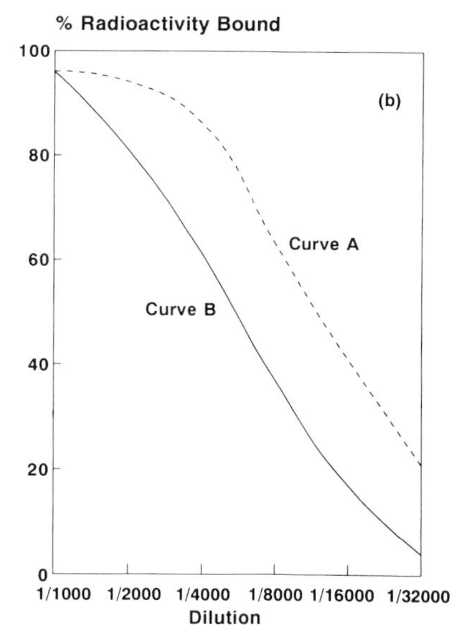

***FIGURE 4.2:** Assessment of tracer damage to radiolabeled antigens. Curve A is obtained with fully radiolabeled antigen (say 1 ng/ml) while curve B is obtained with partially radiolabeled antigen (say 0.1 ng/ml radiolabeled and 0.9 ng/ml unlabeled). An undamaged radiolabeled antigen retaining full immunological activity is shown in **(a)** whilst tracer damage to the antigen is shown in **(b)**.*

A variety of methods are available to assess the quality of labeled antigens used in radioimmunoassays. These include comparisons of the electrophoretic mobility of the labeled and unlabeled antigen and estimation of tracer binding to excess antibody. However, the most stringent test of tracer quality is a direct comparison of the immunoreactivity of the labeled and unlabeled antigens [3]. This can be achieved by determining antiserum titration curves of per cent tracer bound versus antiserum dilution when the tracer is either fully radiolabeled or only partially radiolabeled. If the labeled antigen shows similar immunoreactivity to the unlabeled material, the curves should be superimposable for a given concentration of antigens (see *Figure 4.2a*). If the curves are not superimposable, tracer damage has occurred (see *Figure 4.2b*).

4.1.3 Production of antibody

The performance of a radioimmunoassay depends not only on the quality of the radioantigen, but also on the properties of the antibody. Traditionally, antibodies were produced by immunization of an intact animal. The resultant antiserum contains many slighty different antibodies each directed towards the original antigen; so-called 'polyclonal' antibodies. Recently, the introduction of hybridoma techniques has allowed the selection and culture of cells producing a single antibody type; so-called 'monoclonal' antibodies. Because monoclonal antibodies bind more specifically and often with higher affinity to the original antigen, they are becoming increasingly popular in radioimmunoassay.

In general, if a protein is of sufficiently large molecular mass to be antigenic (see Section 4.1.1), polyclonal antibodies can be produced by injecting small amounts (often as little as 100 μg) into any animal other than that from which the protein was derived. Most laboratories use rabbits for antiserum production whilst sheep, goats, and donkeys are used for large scale commercial production. The protein is usually injected subcutaneously in 1 ml of saline solution at up to four sites. If the protein is strongly antigenic, antibodies can be detected (by Ouchterlony immunodiffusion, see [4]) after approximately 10 days, reach a maximum after a further 5–10 days and then decline. At a suitable time after immunization, up to 15% of total blood volume is withdrawn from a superficial vein; in rabbits the posterior marginal ear vein is invariably used. (Total blood volume is assumed to be 65 ml/kg body weight.) The blood is allowed to clot at 37°C and the serum separated by centrifugation. Proteolytic enzymes and the complement proteins are inactivated by heating the serum to 56°C for 45 min (antibodies are not denatured at this temperature), and the serum is then stored in multiple aliquots at –20 or –70°C. Booster doses of antigen can be given at, say, monthly intervals to maintain antibody product.

When raising antibodies to weakly antigenic compounds, exposure of the immune system to the antigen can be increased by administering it in a form which is slowly released. This can be achieved by either mixing the antigen solution in equal parts with Freund's complete adjuvant (a mixture of mineral oil, detergent and killed mycobacteria) or by administering the antigen in a particulate form. However, Freund's complete adjuvant should be used only when necessary and never more than once in the same animal.

Before antisera to small molecular mass haptens such as non-peptide hormones and drugs can be raised, the haptens must be chemically coupled to a large molecular mass antigenic molecule such as a protein or polysaccharide. Because it is readily available and cheap, bovine serum albumin is often used as the antigenic carrier. A peptide bond is usually formed between a carboxyl group on the hapten and the free amino group on the side chain of lysine residues in the antigenic carrier. If the hapten does not contain a carboxyl group, a derivative must first be formed.

Monoclonal antibodies are usually obtained following immunization of mice. The spleens are removed and minced, and a suspension of spleen cells is incubated with myeloma cells from the same strain of mouse in the presence of polyethylene glycol 6000 which promotes cell fusion and hybridization. The cells are then diluted and cultured in a medium which is toxic to the myeloma cells such that only hybridized cells survive. Each culture is tested for antibody activity and a single 'clone' is selected which produces a single species of antibody immunoglobulin. This can be multiplied either in culture or *in vivo* by injection into the peritoneal cavity of mice.

4.1.4 Optimization of antiserum dilution

The optimal dilution of the antiserum is that which will bind 50–70% of the radioantigen in the absence of unlabeled antigen. The antiserum titration curve (see *Figure 4.2*) will not only yield information on the quality of the radiolabeled antigen, it will also allow the dilution of antiserum required in a radioimmunoassay to be determined. A good antiserum with a high antibody titer will often be used in a radioimmunoassay at dilutions in excess of 1:10 000.

4.1.5 Optimization of antibody–antigen binding

In order to develop a sensitive radioimmunoassay, the antibody must be capable of binding the antigen with high affinity. Antibody–antigen binding is reversible and utilizes non-covalent chemical bonding such as electrostatic interactions, hydrogen bonding and van der Waals forces. These types of chemical bonding are sensitive to changes in the pH of the surrounding medium. It is advisable, therefore, to determine the pH dependency of antibody–antigen binding and to fix the assay pH at that giving maximal binding. In addition, different buffer materials can affect binding and it is customary to try several buffer materials at the optimal pH.

Optimization of radioimmunoassays also requires a knowledge of the time taken for antibody–antigen binding to reach equilibrium. This can be determined simply by assaying the proportional binding of a fixed amount of radioantigen to its antibody after various incubation times. Incubation times to achieve equilibrium can vary from 1 to 72 h. Incubations for less than 4 h are routinely carried out at room temperature, but for those radioimmunoassays requiring greater than 4 h to achieve equilibrium, incubations are best carried out at 4°C to minimize possible degradation or denaturation of antibody and/or antigen.

4.1.6 Separation of antibody-bound and free antigen

A crucial step in any radioimmunoassay is the separation of the antibody-bound radioantigen from the free radioantigen, since incomplete separation would lead to loss of precision and sensitivity. A variety of separation methods are available and have been reviewed in reference [5].

At the concentration employed in a radioimmunoassay, the antibody–antigen complex is almost always soluble. Originally, separation techniques utilized differences in the physicochemical properties of the antibody-bound and free antigen. Such methods included electrophoresis in cellulose acetate or polyacrylamide gels and gel exclusion or ion exchange chromatography. However, such methods are time consuming and laborious, particularly if a large number of samples are to be processed.

If the free antigen is of small molecular mass (i.e. a hapten) it is often possible to absorb it on to solid materials such as dextran coated charcoal, silica gel or hydroxyapatite. Centrifugation will then leave the antibody-bound radioantigen in the supernatant and either the pellet or supernatant (or both) can be counted. Alternatively, the antibody-bound antigen can be precipitated non-specifically with salt (e.g. ammonium sulfate) or solvent (e.g. ethanol). In this case the small molecular mass, free radioantigen is left in the supernatant after centrifugation.

If the antigen is a large polypeptide, a particularly useful method of separation is the so called double antibody technique. Here, a second antiserum is used which is raised to the γ-globulin fraction of the animal species in which the primary antiserum was produced. Thus, if the primary antibody was raised in rabbits, injection of rabbit serum into, say, a donkey will cause the donkey to produce antibodies to all rabbit serum proteins including the γ-globulins. In immunological terms, this serum would be known as donkey anti-rabbit serum and such antisera are commercially available. Thus, after binding between the first antibody and antigen has achieved equilibrium, a suitable dilution of the second antiserum is added which will react specifically with the first antibody to form a separable precipitate. A minor problem encountered with double antibody separation systems is the possible interaction of the second antiserum with the γ-globulins present in the primary antiserum causing a reduction in antigen–antibody binding. A further non-specific effect may be the adsorption of radioantigen to protein in the second antiserum giving rise to the inclusion of free radioantigen in the bound fraction. These problems can be overcome by including in the assay buffer a small amount (0.1–1% v/v) of non-immune carrier serum from the species in which the primary antiserum was raised.

Another approach to separating the antibody-bound and free antigen is to immobilize the antibody. Many commercial radioimmunoassay kits are now supplied where the antibody is attached to a particulate solid support such as Sephadex and can be pelleted by centrifugation. An even more ingenious adaptation is to bind the antibody to magnetizable polymer particles; separation is now achieved by standing the assay tubes on a magnetic base. Finally, in some commercial radioimmunoassay kits, the antibody is bound to the side of the assay tubes in such a way that the free radioantigen can be removed by simply decanting and washing.

4.1.7 Sample preparation

Most radioimmunoassays consist of simply mixing the sample containing the antigen, radioantigen and antibody without any advance preparation of the sample. However, there are two situations when some sample preparation is necessary. Firstly, if the assay does not have sufficient sensitivity, the endogenous antigen must be concentrated. Secondly, if the antibody cross-reacts substantially with components other than the antigen in the sample, these must first be separated. These demands can usually be met by a variety of extraction and chromatographic procedures.

An interesting example of this second consideration is in the radioimmunassay of the opiate peptides leu- and met-enkephalin. Antisera to these peptides show substantial cross-reactivity to each other and also to other closely related neuroactive peptides in plasma and cerebrospinal fluid. A method has recently been developed in which these peptides are first separated by high performance liquid chromatography and the fractions are then analyzed for the presence of leu- and met-enkephalin by classical radioimmunoassay [6].

4.1.8 A typical radioimmunoassay: handling the data

To examine data handling, let us consider a typical radioimmunoassay for a peptide hormone utilizing the double antibody separation technique. The initial incubations will consist of buffer, antiserum, normal non-immune serum, radiolabeled antigen (usually 10 000–20 000 d.p.m./tube) and standards/unknowns. It is advisable to run all incubations in at least duplicate. After a suitable incubation time to reach equilibrium binding, the second antiserum is added. Centrifugation and aspiration of the supernatant will leave the antibody-bound radiolabeled hormone in the pellet for counting. Non-specific binding of the radiolabeled hormone to components in the assay other than the antiserum should always be determined by simply incubating buffer, normal non-immune serum and radiolabeled hormone. A good radioimmunoassay will show less than 10% non-specific binding.

There are three general methods of calculating results from radioimmunoassays:

(a) Linear–linear plots. The count rate for each standard is plotted against hormone concentration on linear graph paper to produce a typical curve as in *Figure 4.1a*. Variations of this include plotting per cent bound, free (i.e. counts in the supernatant), per cent free or bound/free versus hormone concentration. If the count rate in the free radiolabeled hormone fraction has not been determined directly, this can be calculated by subtracting antibody-bound counts from total counts.

(b) Linear–log plots. Plotting any of the above against the logarithm of the hormone concentration will invariably produce a sigmoidal calibration curve with a linear portion in the middle (see *Figure 4.3a*).

(c) Log–log plots. It is essential for log–log plotting that all count rates are corrected for non-specific binding (NSB). The NSB-corrected counts for each standard are expressed as a percentage of the NSB-corrected counts for the zero standard and plotted on a log–log scale against hormone concentration. This plot invariably produces a straight line. An

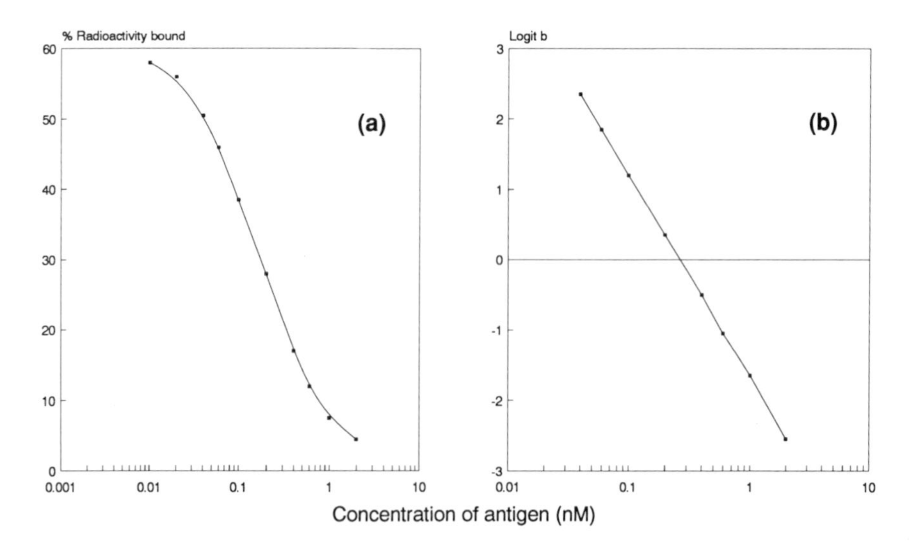

FIGURE 4.3: *Radioimmunoassay calibration curves.* ***(a)*** *Linear–log plot,* ***(b)*** *logit plot where logit* b = *log* b/(100 – b) *and* b *is the proportion of radioantigen bound expressed as a percentage of that in the zero standard.*

alternative is to plot log $b/(100 - b)$ versus log concentration where b = the proportion of radiolabeled hormone bound expressed as a percentage of that in the zero standard; this is known as a logit plot (see *Figure 4.3b*).

Clearly log–log or logit plotting is preferable as this produces a linear assay calibration. This can be fitted easily to a line of best fit for the subsequent determination of unknowns.

4.1.9 Other immunoassays

Instead of attaching an isotopic label to the antigen to be measured, it is also possible to attach an enzyme or fluorescent label. Examples of enzyme labels are alkaline phosphatase and horseradish peroxidase whilst examples of fluorescent labels are fluoescein and rhodamine. The enzyme or fluorescent label is then quantified after separation of antibody-bound and free antigen. These assays are known respectively as enzymoimmunoassays (ELISA – enzyme linked immunosorbent assay) and fluoroimmunoassays (FIA). A full discussion of these immunoassays is clearly beyond the scope of this book (but can be found in reference[7]).

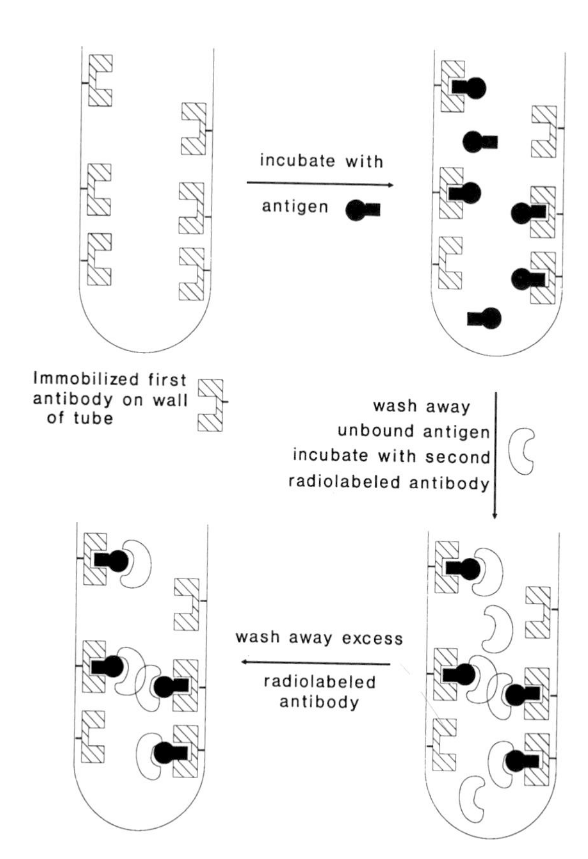

FIGURE 4.4: *Diagrammatic representation of a two-site immunoradiometric assay for estimation of antigens.*

Another variant of the classical radioimmunoassay is to radiolabel the antibody rather than the antigen; this type of assay is known as an immunoradiometric assay (IRMA). Such assays require the use of large quantities of highly specific antibodies and so monoclonal antibodies are invariably used. The most commonly used form of IRMA is the 'two-site' or 'sandwich' assay which utilizes two antibodies, both in excess. The first antibody, which is unlabeled and usually attached to a solid support, is used to bind the antigen. The second antibody is directed towards a different epitope (binding site) on the antigen and is radiolabeled (*Figure 4.4*). Because both antibodies are present in excess, it can be assumed that all of the antigen binds to the first antibody and that one molecule of the radiolabeled second antibody binds to one molecule of antigen. Thus, if the radioactivity in the antibody–antigen complex is determined, there is a direct linear relationship between the bound radiolabeled second antibody and the antigen concentration. This is a major advantage over conventional radioimmunoassay which helps to explain the rapidly increasing popularity of IRMAs.

Two site IRMAs are ideally suited to microtiter plate technology. These plates have many wells (up to 96) which can be coated with the first antibody. After the initial incubation to trap the antigen, a washing stage, a second incubation with the radiolabeled antibody and a final washing stage, the wells can be cut out using a hot wire plate cutter to determine the radioactivity in each well.

4.2 Ligand-binding assays

4.2.1 Definitions

Many biologically active substances (e.g. hormones, neurotransmitters) exert their physiological effects by first binding to receptor molecules. Any substance which binds to a receptor is referred to as a ligand and may be an agonist or an antagonist. Almost all receptors are located in the plasma membrane of target cells; the exceptions are receptors for steroid hormones which are located in the cytosol, and those for thyroid hormones which are bound to the nucleus. Plasma membrane receptors are invariably glycoproteins.

The two most commonly determined parameters in ligand-binding assays are the total available concentration of receptor binding sites (B_{max}) and the equilibrium dissociation constant (K_D). K_D is numerically equal to the ligand concentration necessary to give half-maximal receptor binding and, like the K_M of an enzyme, is effectively a measure of the affinity of a particular ligand for its receptor; the lower the K_D value, the higher the affinity of the ligand for its receptor. K_D and B_{max} are determined graphically from the following equation first suggested by Scatchard [8];

$$\frac{LR}{L} = \frac{LR_{max}}{K_D} - \frac{LR}{K_D}$$

where L = concentration of free ligand, LR = concentration of receptor-bound ligand and LR_{max} = maximum concentration of receptor-bound ligand, that is, B_{max}. Thus, a plot of LR/L versus LR (colloquially known as bound/free versus bound) will produce a straight line where the gradient is equal to $-1/K_D$ and the intercept on the LR axis is equal to LR_{max} (i.e. B_{max}). A typical Scatchard plot is shown in *Figure 4.5*.

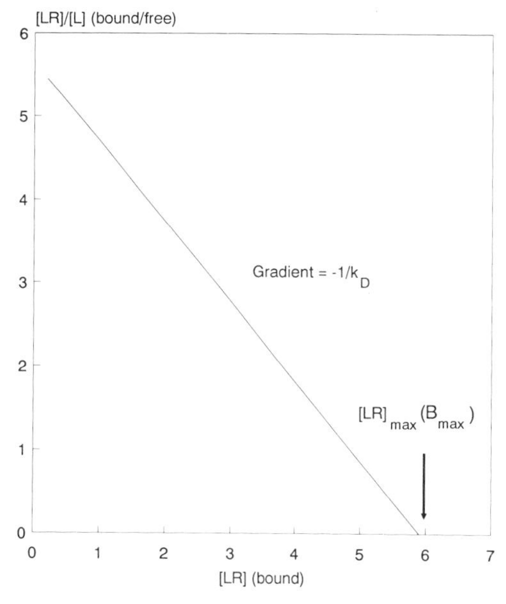

***FIGURE 4.5:** A typical Scatchard plot.*

4.2.2 Assay strategy

Ligand-binding assays depend upon the binding of a radiolabeled ligand to a population of receptors. The receptors may be present on the surface of intact cells, contained within a tissue homogenate or a purified plasma membrane preparation, or, exist in free solution. Thus, a typical ligand-binding assay will involve incubating different concentrations of radiolabeled ligand with a fixed concentration of receptors in a suitable buffer until equilibrium binding is achieved. The receptor-bound radiolabeled ligand is separated from the free radiolabeled ligand, counted and K_D and B_{max} determined by plotting LR/L versus LR. Such an approach can be used to compare receptors in different tissues and to assess the effects of various disease states or drugs upon receptor affinity and concentration. Alternatively, ligand binding assays can be used to investigate the binding of candidate drugs to a wide range of receptors.

An obvious similarity between ligand-binding assays and radioimmunoassays is that, since the reaction between ligand and receptor is a simple reversible bimolecular binding event, it is imperative that assays are incubated for long enough to allow binding to come to equilibrium. Ligand-binding assays are invariably carried out at 0°C to prevent possible degradation of radioligand and/or receptor. Another important consideration is that, if the source of receptor is intact cells and the receptor endocytoses (internalizes) after ligand binding, it is imperative to perform binding assays at 0°C to arrest endocytosis. Such receptors often recycle back to the plasma membrane and recycling (intracellular) receptors can be determined by first permeabilizing intact cells with a low concentration of a detergent such as digitonin [9].

Assays should be carried out at the pH giving optimal binding. In practice, many ligand-binding assays are carried out at pH 7.4 in a buffer of similar composition to extracellular fluid since this simulates the normal environment of the receptor binding site. If the receptor preparation is likely to contain enzymes which degrade the ligand, inhibitors of these enzymes should be included. Catecholamine ligands, for example, are degraded by the membrane bound enzyme monoamine oxidase. Thus, it is customary to include in the assay buffer a monoamine oxidase inhibitor such as parglyine together with an antioxidant (e.g. dithiothreitol or ascorbic acid) to inhibit chemical oxidation of catecholamines [10].

4.2.3 Choice of radioisotope

^{3}H-labeling is most commonly used for most small molecular mass ligands and ^{125}I-labeling for peptide ligands. Many receptors are present at extremely low concentrations (20–200 fmol/mg of membrane protein) and in order to achieve detectable levels of receptor-bound radioactivity, the radiolabeled ligand must have a specific activity of at least 185 GBq/mmol [11]. Whilst this is easily obtainable with ^{3}H- or ^{125}I-labeling, it cannot be achieved with ^{14}C-labeling (see Section 2.2). It is also important that the radiolabeled and unlabeled ligand should have identical affinity for the receptor. Whilst tracer damage is extremely unlikely with ^{3}H-labeling, radioiodination of tyrosine residues by the chloramine T or lactoperoxidase methods (see Section 2.3.1) can affect the structure of peptides.

Many peptides contain more than one available tyrosine residue and radioiodination will produce a mixture of mono- and multi-iodinated products. It is preferable to use monoiodinated peptides in ligand-binding assays since these are least likely to suffer from tracer damage. Monoiodinated products can be separated from multi-iodinated products by traditional ion exchange chromatography or by reverse phase HPLC. Insulin, for example, contains four tyrosine residues capable of radioiodination; these are at positions 14 and 19 in the 21 amino acid A

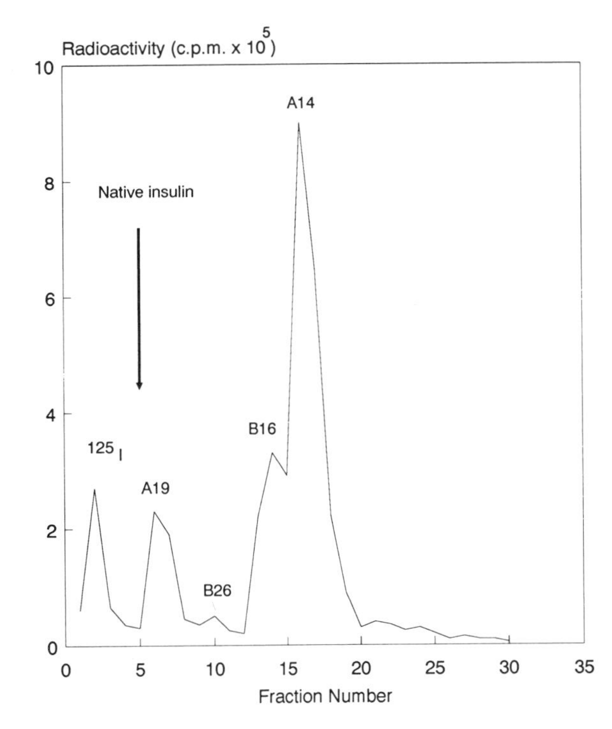

FIGURE 4.6: *Preparation of [A14-^{125}I]-monoiodoinsulin by reverse phase HPLC. Iodinated insulins are separated isocratically with 26.5% acetonitrile on 0.1 M ammonium acetate buffer, pH 5.5. By kind permission of Drs A.P. Bevan and G.D. Smith.*

chain and at positions 16 and 26 in the 30 amino acid B chain. A14-Monoiodoinsulin can now be prepared free of other iodoinsulins by reverse phase HPLC (see *Figure 4.6*), and, since it shows similar biological activity and similar affinity for the insulin receptor compared to the native unlabeled hormone, it is most frequently used in studies involving insulin interactions with its receptor [12]. However, if the peptide contains many tyrosine residues, it becomes impossible to separate the vast number of resultant iodopeptides and the radiolabeled product is used unfractionated. If tracer damage of iodopeptides is suspected, it is sometimes better to radioiodinate the peptide by the Bolton and Hunter method (see Section 2.3.2) since it is less damaging to the peptide.

4.2.4 Separation of bound and free ligand

A variety of methods are available to separate the receptor-bound and free radioligand in ligand-binding assays. Broadly speaking, these are similar to those used to separate antibody-bound and free antigen in radioimmunoassays. These methods fall into two groups depending upon whether the receptor population is particulate (i.e. membrane bound) or soluble.

If the receptor is particulate, receptor-bound radioligand can be harvested on filter discs following vacuum filtration. Vacuum filtration manifolds are available commercially and can handle several samples at once. Alternatively, receptor-bound radioligand can simply be pelleted by centrifugation and the free radioligand in the supernatant decanted. This is particularly useful when performing ligand-binding assays on intact cells since these will pellet quickly at low *g*-forces. A refinement of this is to pellet the cells (or membranes) through a cushion of silicone oil at the bottom of the centrifuge tube; this will minimize contamination of the receptor-bound radioligand (pellet) with free radioligand (supernatant) [13].

If the receptor is soluble and the radioligand of small molecular mass it may be possible to absorb non-specifically the radioligand on to charcoal or silica gel; centrifugation will now leave the receptor-bound radioligand in the supernatant. Another approach is to precipitate the receptor-bound radioligand with ammonium sulfate or ethanol which can be removed in the pellet after centrifugation. Gel exclusion chromatography can also be used to separate receptor-bound and free radioligand but would be time consuming with many samples. Finally, a more complex separation system is equilibrium dialysis [14]. The incubation containing both receptor-bound and free radioligand is placed on one side of a dialysis membrane of molecular weight cut-off such that the free radioligand can diffuse to achieve an equal concentration on the other side of the membrane whilst the receptor-bound radioligand cannot (see *Figure 4.7*).

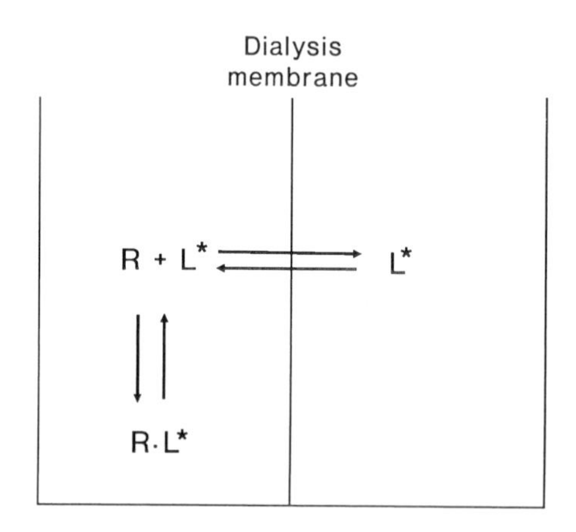

FIGURE 4.7: *Separation of receptor-bound radioligand (R.L*) from free radioligand (L*) by equilibrium dialysis. The dialysis membrane has a molecular weight cut-off such that R.L* remains on the left side whilst L* is free to diffuse to the right side.*

Radioactivity measurements on the two compartments will give the amount of free radioligand and the combined amount of free and receptor-bound radioligand. The amount of receptor-bound radioligand can then be calculated by difference.

4.2.5 Non-specific binding

A major problem with some ligand-binding assays is that the radioligand binds with low affinity to sites other than receptors in the assay mixture;

this is known as non-specific binding. In general, the more impure the receptor population, the more likely it is that appreciable non-specific binding will occur. Thus, non-specific binding is usually substantial when using intact cells, homogenates and, to a lesser extent, plasma membrane preparations, whilst it is often insignificant when using purified, soluble receptors. Non-specific binding of radioligand increases linearly with

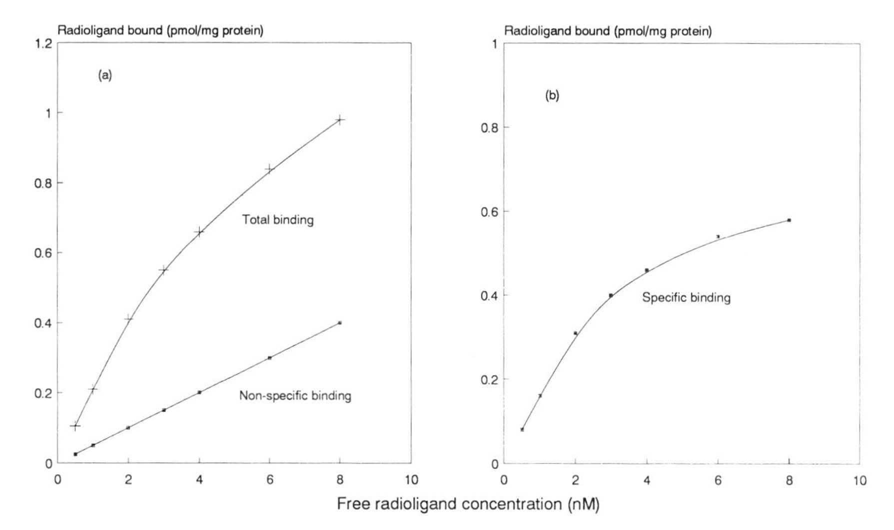

FIGURE 4.8: *Non-specific binding in ligand-binding assays.* ***(a)*** *Total binding (+) and non-specific binding determined in the presence of a saturating excess of unlabeled ligand (•).* ***(b)*** *Specific binding obtained by subtracting non-specific from total binding.*

increasing radioligand concentration whilst specific binding of radioligand and receptor is saturable at the B_{max} value (see *Figure 4.8a*). Thus, non-specific binding is determined as radioligand binding in the presence of a saturating excess of unlabeled ligand (usually at least 100 times the K_D value). Specific binding of radioligand is determined by subtracting the non-specific binding from total binding (see *Figure 4.8b*).

4.2.6 Calculation of K_D and B_{max}

In order to illustrate this, let us consider a typical binding assay in which various concentrations of [A14-^{125}I]insulin were incubated with 20 μg of adipose tissue plasma membrane protein in a final volume of 1 ml of buffer. The specific activity of the [A14-^{125}I]insulin was 10^4 c.p.m./pmol. After incubation overnight at 0°C (during which time binding equilibrium was reached) receptor-bound insulin was harvested by vacuum filtration on filter discs presoaked in buffer. The filter discs were counted for radioactivity in a gamma counter. Non-specific binding was determined

TABLE 4.1: *Binding of [A14-^{125}I] insulin to adipocyte plasma membranes*

Total insulin (nM)	Bound insulin (c.p.m.)	Bound c.p.m. in presence of 200 nM unlabeled insulin
0.4	1800	400
1.0	3780	980
1.6	5340	1500
2.5	7430	2570
5.0	10 730	4910
10.0	17 020	10 400

by carrying out similar incubations in the presence of 200 nM unlabeled insulin. The results are given in *Table 4.1*, their analysis for Scatchard plotting is shown in *Table 4.2* and the resultant Scatchard plot is shown in *Figure 4.9*.

TABLE 4.2: *Processing of [A14-^{125}I] insulin receptor binding data for Scatchard plotting*

(a) Specific binding of [A14-^{125}I] insulin	(b) Bound insulin (pmol)	(c) Bound (pmol/mg of membr-ane protein)	(d) Free insulin (nM)	(e) Bound/free (pmol/mg/nM)
1400	0.14	7.0	0.26	26.9
2800	0.28	14.0	0.72	19.4
3840	0.384	19.2	1.216	15.8
4860	0.486	24.3	2.014	12.1
5820	0.582	29.1	4.418	6.59
6620	0.662	33.1	9.338	3.54

Specific binding of [A14-^{125}I] insulin is calculated by subtracting non-specific from total binding (column a). Knowing the specific activity of the [A14-^{125}I] insulin used, this can be converted to pmol of bound insulin (column b) which in turn is expressed as pmol/ mg of membrane protein (column c). The concentration of free [A14-^{125}I] insulin is calculated by subtracting the receptor- bound radioligand from its total concentration in each assay tube (column d) and bound/free is now simply calculated (column e).

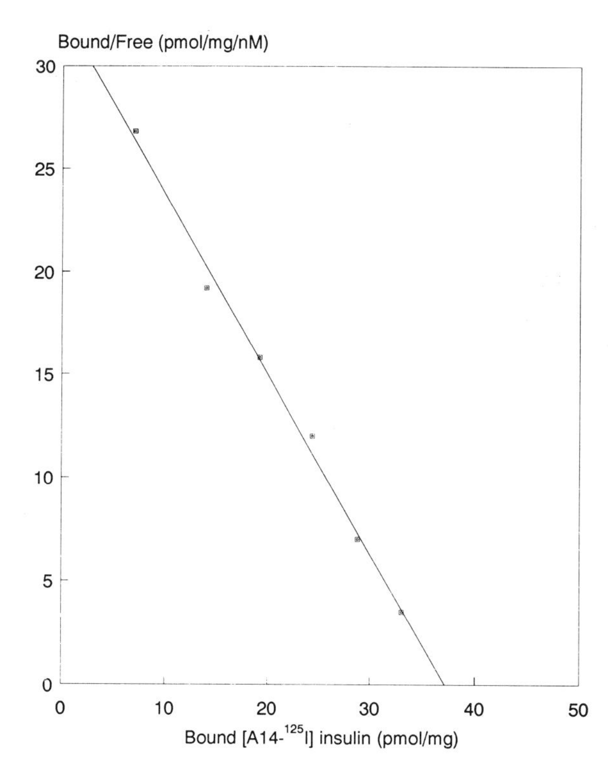

FIGURE 4.9: *Scatchard plot of the binding of [A14-^{125}I] insulin to adipocyte plasma membranes. The gradient of the line (0.847 nM^{-1}) is equal to $-1/K_D$ giving a K_D of 1.18 nM, and the intercept on the bound axis is equal to B_{max} (37.5 pmol/mg of membrane protein).*

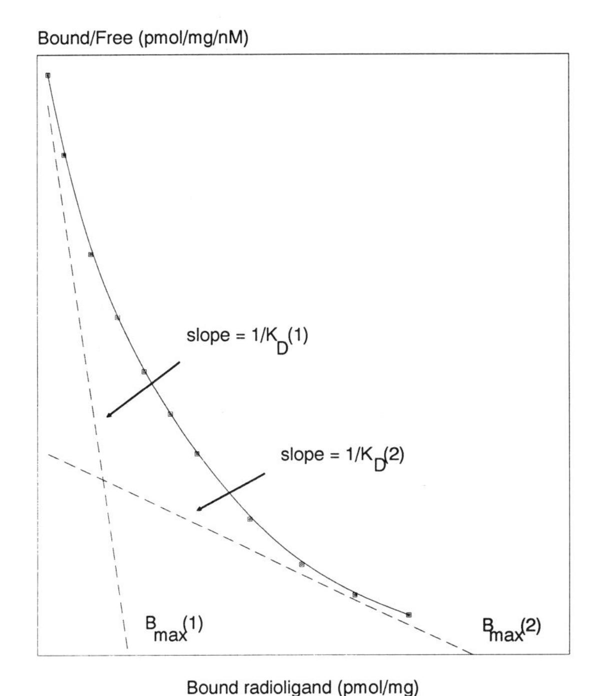

FIGURE 4.10: *Non-linear Scatchard plot due to two distinct receptor binding sites. The curve can be resolved into high- and low-affinity binding sites.*

In some cases, ligand-binding assays produce non-linear Scatchard plots as shown in *Figure 4.10*. One explanation for such a plot is negative cooperativity where the affinity for the radioligand decreases as the receptor binding sites become increasingly saturated. An alternative explanation is that the radioligand is binding to a heterogeneous population of sites, that is, at least two functionally distinct receptors are present. Such plots can be resolved by computer non-linear least squares curve fitting programs to produce K_D and B_{max} values for both the high- and low-affinity binding sites (see *Figure 4.10* and [15]).

References

1. Yallow, R.S. and Berson, S.A. (1960) *J. Clin. Invest.,* **39**, 1157–1175.
2. Ekins, R.P. (1976) in *Hormone Assays and Their Clinical Application* (J.A. Lorraine and E.T. Bell, eds). Churchill-Livingstone, Edinburgh, p. 1–72.
3. Bolton, A.E. (1985) *Radioiodination Techniques,* Review 18. Amersham International, UK.
4. Munro, A.C., Chapman, R.S., Templeton, J.G. and Fatori, D. (1983) in *Immunoassays for Clinical Chemistry* (W.M. Hunter and J.E.T. Corrie, eds). Churchill-Livingstone, Edinburgh, p. 447–456.
5. Chard, T. (1987) *An Introduction to Radioimmunoassay and Related Techniques.* Elsevier, Amsterdam.
6. Venn, R.S. (1987) *J. Chromatogr.,* **423**, 93–104.
7. Nakamura, R.M., Voller, A. and Bidwell, D.E. (1986) in *Handbook of Experimental Immunology, Vol. 1, Immunochemistry* (D.M. Wier, ed.). Blackwell Scientific Publications, Oxford, p. 27.1–27.20.
8. Klotz, I.M. (1983) *Trends Pharmacol. Sci.,* **4**, 253–255.
9. Weigel, P.H. and Oka, J.A. (1983) *J. Biol. Chem.,* **258**, 5095–5102.
10. Withy, R.M., Mayer, R.J. and Strange, P.G. (1980) *FEBS Lett.,* **112**, 293–295.
11. Strange, P.G. (1988) in *Radiochemicals in Biomedical Research* (E.A. Evans and K.G. Oldham, eds). Wiley, Chichester, p. 56–93.
12. Evans, E.A. and Oldham, K.G. (1988) in *Radiochemicals in Biomedical Research* (E.A. Evans and K.G. Oldham, eds). Wiley, Chichester, p. 1–13.
13. Minuk, G.Y., Vergalla, J., Ferenci, P. and Jones, E.A. (1984) *Hepatology,* **4**, 180–185.
14. Englund, P.T., Huberman, J.A., Jovin, T.M. and Kornberg, A. (1969). *J. Biol. Chem.,* **244**, 3038–3044.
15. Bürgisser, E. (1984) *Trends Pharmacol. Sci.,* **5**, 142–144.

5 Radioisotopes and Tracer Techniques

5.1 Designing a radiotracer experiment

A major use of radioisotopes is as tracers of metabolic and transport processes. Such experiments are not simply a matter of adding a radiolabeled compound to a biological system and following the radiolabel through the metabolic or transport process. Many factors must be taken into account in designing such experiments.

5.1.1 Choice of radioisotope

A first consideration is to decide which radioisotope to use. The most commonly used isotope for metabolic studies of small molecules is ^{14}C. This allows the metabolic fate of the carbon skeleton of compounds to be determined and, if the compound is oxidized, this can be monitored by following $^{14}CO_2$ release. One of the basic assumptions of using radiolabeled compounds as tracers is that the chemical amount used does not make a significant contribution to the pool size of the compound. However, if the compound being studied occurs at low endogenous concentrations, the maximum specific activity achievable with one ^{14}C per molecule (2.31 GBq/mmol) may not be sufficient. In such cases ^{3}H-labeling may be necessary to produce higher specific activities. However, some enzyme catalyzed reactions involve hydrogen exchange between the compound and cellular water and this can be mis-interpreted as metabolic utilization of the compound. Nevertheless, this has been used to advantage in some applications and the use of ^{3}H-labeled glucose in determining the kinetics of glucose metabolism *in vivo* is discussed in Section 5.4.1.

Other radioisotopes are used to trace the fate of atoms other than C and H in molecules; for example, ^{32}P is often used to investigate the metabolism of nucleotides and phospholipids, and ^{125}I is invariably used to radiolabel proteins in metabolic studies.

5.1.2 Intramolecular position of the radiolabel

If the incorporation of a metabolite into other molecules, or its oxidation to CO_2, is being studied, it is often most suitable to use compounds uniformly labeled with ^{14}C, for example, [U-^{14}C]leucine for studies of amino acid incorporation into protein or [U-^{14}C]glucose for studies of glucose oxidation. For more precise studies of metabolic pathways, the radiolabel must be located in specific chemical groups. An interesting example of this is the anti-inflammatory drug eterylate (*Figure 5.1*). This compound contains two pharmacologically active groups, salicylic acid and *p*-acetamidophenol, linked by an ethylene bridge between carboxylic and phenolic functions. A major metabolic fate of eterylate in mammals is hydrolysis of the ester bond to yield the two pharmacologically active groups (*Figure 5.1*). Thus, in order to study the metabolic fate of the salicylic acid and *p*-acetamidophenoxyacetic acid groups, the parent molecule must be radiolabeled in both groups. In fact, metabolic studies were carried out using the molecule radiolabeled with ^{14}C in the carboxyl function of the salicylic acid group and uniformly in the aromatic ring of the *p*-acetamidophenol group (*Figure 5.1*) [1].

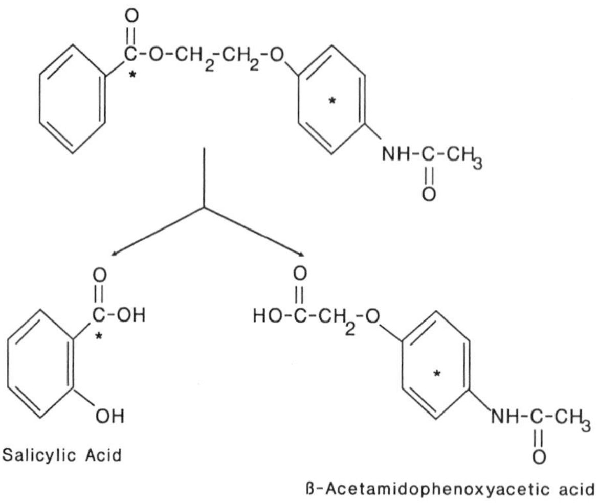

FIGURE 5.1: *Metabolism of the anti-inflammatory drug, eterylate. In order to study the metabolic fates of the hydroysis products, the molecule was labeled with ^{14}C in both the carboxyl function of the salicylic acid group and uniformly in the aromatic ring of the* p*-acetamidophenol group (indicated by *).*

Having decided which isotope to use, and its position in the tracer molecule, an obvious consideration is whether the radiolabeled compound is commercially available or has to be synthesized. It is also important to estimate the amount required in any experiment. This should be as low as will allow accurate determination by the proposed detection system (machine counting or autoradiography). An excess factor of 2–4-fold should be incorporated to allow for unexpected dilution factors or losses.

5.1.3 Isotope effects

A basic assumption when using radioisotopes as tracers is that the radiolabeled substance behaves chemically and physiologically exactly like the mother substance. In this respect, extra care should be taken when using isotopes of hydrogen in metabolic studies because of the so-called isotope effect. An isotope effect is defined as any difference in the chemical or physical behavior of two compounds which differ only in the isotopic composition of one (or more) of their chemical elements. In biological systems, by far the most important isotope effects are on rates of reactions in successive steps. Deuterium atoms (^{2}H) are twice as heavy as hydrogen atoms (^{1}H) and therefore the C—D bond has a lower frequency of vibration than the C—H bond. As a result, breakage of a C—D bond requires a higher activation energy than a C—H bond and several-fold differences in rate have been observed. For example, yeast alcohol dehydrogenase has a V_{max} for isotopically normal ethanol which is 1.8 times greater than that for deuteriated ethanol [2]. Even greater differences are possible for tritium (^{3}H). Thus, if a metabolic study involves monitoring the rupture of C—H bonds (as occurs, for example, in hydroxylation reactions catalyzed by mixed function oxidases), the use of ^{2}H or ^{3}H to monitor the rate of this process can give misleading results. For isotopes of higher atomic mass, the isotope effect is far less serious because the mass difference is much less. For example, ^{14}C is only 17% heavier than ^{12}C and rate differences of only approximately 10% occur for carbon bond rupture. Mass differences between ^{32}P and ^{31}P and between ^{35}S and ^{32}S are even smaller.

5.2 Radioisotopic methods of enzyme assay

5.2.1 General principles

Probably the most common strategy for measuring the activity of an enzyme is to utilize the chromogenic properties of a substrate, product or cofactor. If none of these molecules is easy to measure spectrophotometrically, the enzymic reaction is often linked to a second reaction to produce a useful chromogen. This second reaction may be purely chemical or itself may be enzymic.

Radioisotopic methods have also been developed for certain enzymes. Sometimes they are merely alternatives to spectrophotometric assays which provide either a more convenient methodology or a greater sensitivity. However, when a useful chromogen is not available, they provide the only useful technique. If the enzyme reaction is simple, either utilization of radiolabeled substrate or formation of radiolabeled product may be monitored. However, for a two stage enzymic reaction in which intermediates accumulate, only substrate utilization will measure the true rate. This condition also applies when the reaction product(s) are further utilized by other enzymes in a metabolic pathway.

Thus, a typical radioisotopic enzyme assay involves incubating the radiolabeled substrate with the enzyme under optimal conditions (temperature, pH, substrate and cofactor concentration, allosteric activators, etc.) for a fixed period of time. The residual radiolabeled substrate is then separated from the radiolabeled product(s) prior to determination of radioactivity in either substrate or product(s). The success of any radioisotopic enzyme assay is critically dependent on the complete separation of the radiolabeled substrate and products. Several methods have been used and include precipitation, solvent extraction, ion exchange and paper chromatography, electrophoresis and charcoal adsorption. It is imperative that the reaction rate remains constant throughout the incubation period and this should be validated when setting up a radioisotopic enzyme assay. A further useful validation is that the reaction rate should show a linear relationship with the enzyme concentration.

5.2.2 Decarboxylases

One of the classic radioisotopic enzyme assays is for decarboxylating enzymes [3]. Such assays depend on the release of $^{14}CO_2$ from a substrate

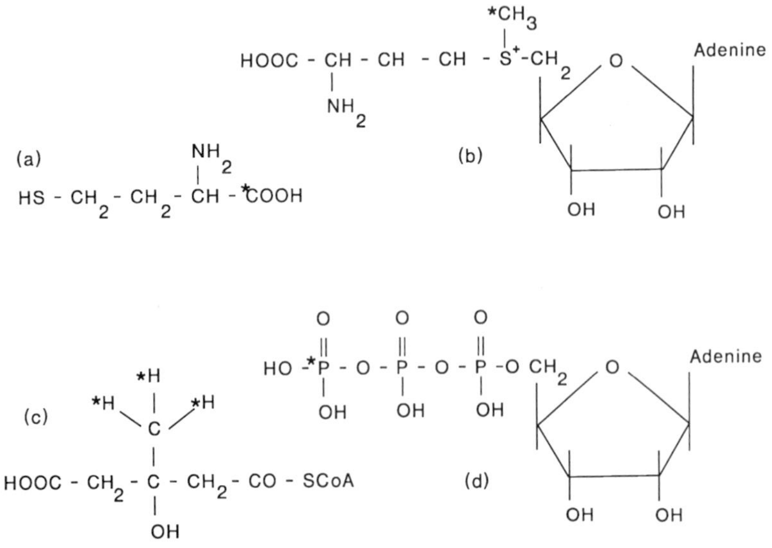

(a) $HS-CH_2-CH_2-CH(NH_2)-{}^{*}COOH$

(b) $HOOC-CH(NH_2)-CH-CH-S^{+}({}^{*}CH_3)-CH_2$–ribose(OH, OH)–Adenine

(c) $HOOC-CH_2-C(OH)(C{}^{*}H_3)-CH_2-CO-SCoA$

(d) $HO-{}^{*}P(=O)(OH)-O-P(=O)(OH)-O-P(=O)(OH)-O\,CH_2$–ribose(OH, OH)–Adenine

FIGURE 5.2: *Structures of some radiolabeled substrates used in enzyme assays (* indicates the position of radioactive atoms).(**a**) [1-^{14}C]methionine used to assay amino acid decarboxylation. (**b**) S-adenosyl-L-[methyl-^{14}C]methionine used in the assay of methyl transferases. (**c**) [3-^{3}H]HMG-CoA used in the assay of HMG-CoA reductase. (**d**) [γ-^{32}P]ATP used in the assay of protein kinases.*

radiolabeled with ^{14}C in the carboxyl group destined for removal (e.g. [1-^{14}C]methionine for amino acid decarboxylation, see *Figure 5.2*). The ^{14}C-labeled substrate is incubated with the enzyme in a container made gas tight with a rubber seal; manometer flasks with a center well and a single side-arm are ideal for this purpose (see *Figure 5.3*). At physiological pH, the ^{14}C is first released as $H^{14}CO_3^-$ into the buffer. This can be liberated as $^{14}CO_2$ by acidification; either acid can be tipped into the main compartment of the manometer flask from the side arm or injected through the rubber seal. The released $^{14}CO_2$ is trapped on filter discs in the center well soaked in either alkali (e.g. KOH) or an amine (e.g. 3-phenylethylamine), and then counted. Unsatisfactory results are usually due to either incomplete acidification leading to incomplete liberation of $^{14}CO_2$, or insufficient time being allowed for trapping all the liberated $^{14}CO_2$.

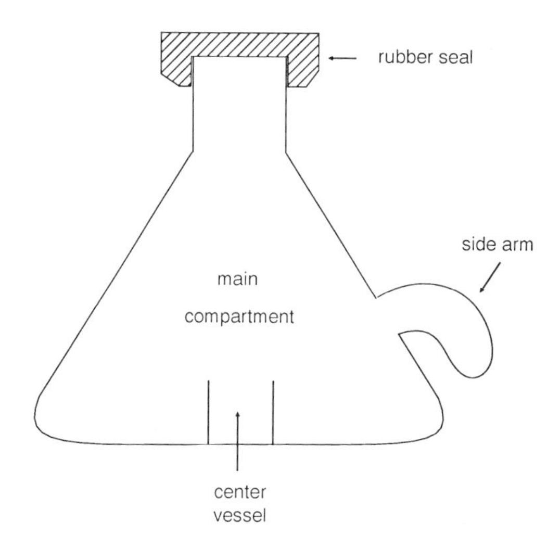

***FIGURE 5.3:** A typical manometer flask used in experiments where trapping of $^{14}CO_2$ is necessary.*

5.2.3 Transferases

Many methyl transferring enzymes have been assayed using the methyl donor labeled with ^{14}C in the methyl group destined for transfer to another molecule [4]. For example, *S*-adenosyl-L-[methyl-^{14}C]methionine (*Figure 5.2*) has been used in the assay of histamine-*N*-methyltransferase and hydroxyindole-*O*-methyltransferase, and [5-^{14}C]methyl-tetrahydrofolic acid in the assay of N^5-methyltetrahydrofolate-homocysteine transmethylase. The ^{14}C-labeled methylated products are separated from the radiolabeled substrates by differential extraction.

Sugar transferases can be assayed by following the incorporation of ^{14}C-labeled sugars into suitable acceptor molecules. For example, galactosyl transferase can be assayed by incubating the enzyme with UDP-[U-

^{14}C]galactose and *N*-acetyl glucosamine as acceptor [5]. After stopping the reaction with ethylenediaminetetraacetic acid (EDTA) and neutralizing to pH 7.4, unreacted UDP-[U-^{14}C]galactose is separated from the product (lactosamine) by ion exchange chromatography.

5.2.4 Cholesterol metabolism

The regulatory enzymes of cholesterol metabolism are attracting much attention as possible sites of pharmacological intervention in hypercholesterolemia (a condition which predisposes people to coronary heart disease). These enzymes are invariably assayed using radiochemical substrates. For example, 3-hydroxy-3-methylglutaryl-CoA (HMG-CoA) reductase, the rate limiting enzyme of cholesterol synthesis, is routinely assayed by the conversion of [3-^{3}H]- or [3-^{14}C]HMG-CoA (*Figure 5.2*) to mevalonic acid [6]. The deproteinized incubation mixture is added directly to a toluene-based scintillation cocktail. The reaction product, [^{3}H]- or [^{14}C]mevalonic acid, partitions into the toluene while unreacted [^{3}H]- or [^{14}C]HMG-CoA remains in the aqueous phase and is not detected on scintillation counting. Similarly, cholesterol-7α-hydroxylase, the rate limiting enzyme in the conversion of cholesterol to bile acids, can be assayed by measuring the synthesis of 7α-hydroxycholesterol from any ^{14}C-labeled cholesterol [7]. 7α-Hydroxycholesterol is separated from unreacted cholesterol by thin layer chromatography and the radiolabeled product quantified by either autoradiography of the chromatogram or scraping the spots from the chromatogram and scintillation counting. Acyl-CoA:cholesterol acyltransferase (ACAT), an intracellular enzyme which synthesizes cholesterol esters from cholesterol and a long-chain acyl-CoA, is assayed by following the incorporation of [1-^{14}C]oleyl-CoA into the cholesterol ester fraction [8]. The reaction is stopped by addition of solvent and the radiolabeled lipids extracted into chloroform. Radiolabeled cholesterol esters are separated from other lipids by thin layer chromatography prior to either autoradiography or scintillation counting.

5.2.5 Protein kinases

Many hormones and neurotransmitters exert their effects via phosphorylation of intracellular effector proteins which alters their biological activity. Several hormone-stimulated protein kinases are known. The cyclic nucleotide-dependent protein kinases phosphorylate predominantly at serine residues and are stimulated by either cyclic AMP or cyclic GMP. Ca^{2+}-dependent protein kinases and protein kinase C (which is activated by diacylglycerol) phosphorylate at both serine and threonine residues. Tyrosine kinases, which are stimulated by insulin and some growth factors, phosphorylate at tyrosine residues. The detection and assay of these protein kinase enzymes is critical to investigating the mechanism of action of many hormones and neurotransmitters. Since protein kinases transfer the phosphoryl group from the γ-position of ATP to either serine, threonine or tyrosine residues, they can be detected by following the incorporation of ^{32}P into proteins after incubation with [γ-^{32}P]ATP (*Figure 5.2*) [9]. After terminating the incubation, proteins in the incubation mixture are separated by sodium dodecyl sulfate polyacrylamide gel electrophoresis and ^{32}P-labeled proteins visualized by autoradiography. The phosphate group of phosphotyrosine is resistant to alkaline hydrolysis whilst that of phosphoserine and phosphothreonine is alkaline labile [10]. Thus, comparison of control and alkaline-treated gels allows assessment of whether phosphorylation was at tyrosine or serine/threonine residues. Alternatively, individual phosphoamino acids can be determined by either thin layer chromatography, HPLC or paper electrophoresis in acid hydrolysates of phosphoproteins.

5.2.6 Phosphatases

A number of radioisotopic assays for phosphatases are in common use, for example 5'-nucleotidase (which catalyzes the hydrolysis of AMP to adenosine and inorganic phosphate) and Na^+/K^+-stimulated Mg^{2+}-ATPase. 5'-Nucleotidase can be assayed by incubation of AMP radiolabeled with ^3H or ^{14}C in any position(s) of the adenosine moiety; residual ^3H- or ^{14}C-labeled AMP is precipitated with $ZnSO_4/Ba(OH)_2$ [11]. Similarly, Na^+/K^+-stimulated Mg^{2+}-ATPase can be assayed by incubation with [γ-^{32}P]ATP (*Figure 5.2*); residual [γ-^{32}P]ATP is removed by selective adsorption onto charcoal [11]. In each case, the radioactivity in the liquid phase is a measure of the enzyme activity.

These methods improve the sensitivity of the enzyme assay and eliminate the need to measure inorganic phosphate spectrophotometrically. Although chemical methods for phosphate determination are simple to carry out, they are notoriously prone to contamination from the sample or glassware. Moreover, the contribution to the hydrolysis from the activity of non-specific phosphatases can be avoided by the inclusion of an excess of a second non-radioactive substrate, for example, β-glycerophosphate in the 5'-nucleotidase assay.

5.3 Use of radioisotopes to investigate metabolic pathways

5.3.1 Concepts

Radioisotopes are often used for tracing metabolic pathways. Such experiments usually involve pulse-chase techniques where the radioactive substance is presented to the biological system as a pulse and the radiolabel chased through various metabolites or cellular compartments.

For metabolic studies, samples are taken at various times and the radiolabeled metabolites are separated by chromatographic techniques. These metabolites can often be identified by comparision of their chromatographic behavior to authentic reference compounds although mass spectrometry is often necessary to elucidate their structure. Thus, by comparing the time course of appearance of radiolabeled metabolites, the sequence of metabolites in a metabolic pathway can often be elucidated.

Tracers in metabolic studies are invariably radiolabeled in one or more specific positions in the molecule. Chemical degradation of the radiolabeled metabolites can be used to locate the specific position(s) of the radiolabel in the metabolites. Such an approach not only gives information concerning the path of the radiolabel through a metabolic pathway, but often gives information concerning reaction mechanisms in the pathway.

5.3.2 The TCA cycle

Although first elucidated without the aid of radiolabeled tracers [12], the TCA cycle pathway has subsequently been confirmed by the use of acetate radiolabeled with ^{14}C in either the carboxyl or methyl groups (that is, [1-^{14}C]- or [2-^{14}C]acetate).

When either [1-^{14}C]- or [2-^{14}C]acetate is provided to metabolically active

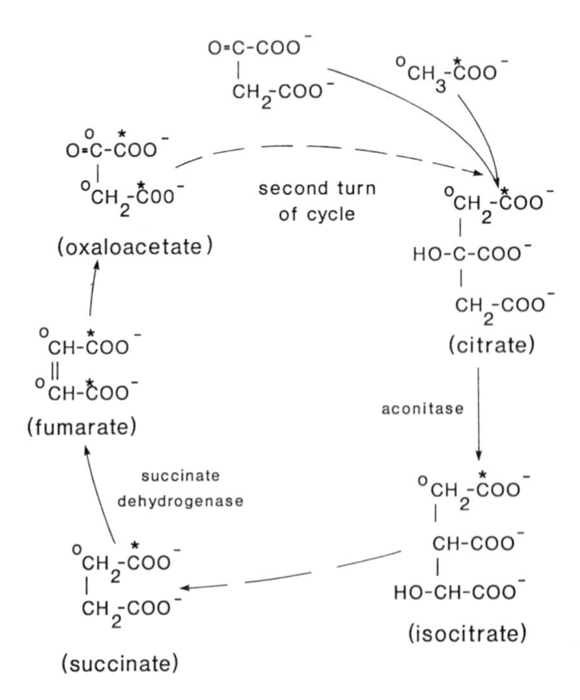

*FIGURE 5.4: Radiolabeling pattern after entry of ^{14}C-labeled acetate into the TCA cycle. *C represents the carboxyl carbon (i.e. [1-^{14}C]acetate), °C represents the methyl carbon (i.e. [2-^{14}C]acetate).*

cells or tissues, this is incorporated into citrate (see *Figure 5.4*). Because citrate is a symmetrical molecule (i.e. it does not have an asymmetric carbon atom), the next enzyme of the cycle, aconitase, would be expected to isomerize the hydroxyl group to either the 'top' or 'bottom' of the isocitrate molecule, thereby randomizing the radiolabel. However, chemical degradation of isocitrate formed from either [1-^{14}C]- or [2-^{14}C]acetate showed that aconitase moved the hydroxyl group away from the newly added ^{14}C atoms. Thus, the use of radiolabeled substrates showed that the enzyme aconitase can act asymmetrically upon citrate, even though citrate is a symmetrical molecule. Succinate, formed later in the TCA cycle, is also a symmetrical molecule and is oxidized to fumarate by succinate dehydrogenase (see *Figure 5.4*). Chemical degradation of fumarate showed that the ^{14}C randomized between the two carboxyl carbons and between the two internal carbons. In other words, succinate dehydrogenase can act upon its substrate symmetrically. Both carboxyl carbon atoms of oxaloacetate are lost as CO_2 in the next turn of the TCA cycle. Thus, $^{14}CO_2$ is not released from [1-^{14}C]acetate until the second turn of the cycle. ^{14}C from [2-^{14}C]acetate persists through a second turn of the cycle and becomes randomized between all four carbon atoms of fumarate. Thus, $^{14}CO_2$ is not released from [2-^{14}C]acetate until a third turn of the cycle and then only half of the ^{14}C is lost.

5.3.3 Regulation of metabolism

The sequence and mechanisms of the reactions of almost all pathways of energy metabolism are now well characterized. However, radiolabeled substrates are still much used to assess the effects of different conditions or regulators upon metabolic pathways. For example, release of $^{14}CO_2$ from ^{14}C-labeled branched chain amino acids has been used to monitor the mitochondrial oxidation of the resultant 2-oxoacid after transamination of the amino acid [13]. Similarly, release of $^{14}CO_2$ from [U-^{14}C]glucose is used to monitor glucose oxidation by various tissues and cells. Measurement of exhaled $^{14}CO_2$ after intravenous administration of ^{14}C-labeled substrates to animals is also possible. However, great care must be taken in interpreting results and the pool size of available substrate must be taken into account. For example, if the pool size of available glucose is decreased to a greater extent than its rate of oxidation, then a net increase in exhaled $^{14}CO_2$ from [U-^{14}C]glucose would be expected [14].

The incorporation of ^{14}C-labeled gluconeogenic precursors (e.g. pyruvate, lactate, alanine, etc.) into cellular glucose and glycogen is often used to monitor hepatic gluconeogenesis [15]. A commonly used gluconeogenic substrate is [1-^{14}C]pyruvate; incorporation of ^{14}C from [1-^{14}C]pyruvate into glucose and glycogen will measure gluconeogenesis. However, pyruvate can also be oxidized to acetyl-CoA by pyruvate dehydrogenase prior to entry into either the TCA cycle or lipogenesis. Thus, simultaneous quantitation of $^{14}CO_2$ release from [1-^{14}C]pyruvate will measure pyruvate oxidation to acetyl-CoA since it is C-1 of pyruvate which is lost in the oxidative decarboxylation reaction catalyzed by pyruvate dehydrogenase.

5.3.4 Drug metabolism

One of the major, present-day uses of radiolabeled compounds in metabolic studies is in preclinical drug development [16]. When a candidate drug is selected for further study on the basis of animal pharmacology experiments, the initial objectives include assessment of possible toxicity, determining the drug's pharmacokinetics (see Section 5.4.2) and evaluating its metabolism. These studies are carried out in experimental animals. The metabolism of candidate radiolabeled drugs is studied from three standpoints:

(a) Excretion studies. These studies are usually carried out with ^{14}C-labeled drugs and give information on the fate of the drug irrespective of its metabolism. Following administration of the ^{14}C-labeled drug to experimental animals, radioactivity excreted in urine and feces is determined. If administered orally, the urinary excretion of ^{14}C provides a minimum value for the amount absorbed. If substantial oxidation of the drug occurs, it is necessary to lead expired air through a CO_2-trap (e.g. KOH or 3-phenylethylamine) to monitor $^{14}CO_2$ production.

(b) Tissue distribution. Quantitative tissue distribution data are best obtained by direct measurement of radioactivity in tissues excized at specific times after dosing with the ^{14}C-labeled drug. Known weights of tissue are either combusted completely such that all radioactivity is converted to $^{14}CO_2$ which can be trapped and counted, or solubilized prior to counting. Alternatively, qualitative or semi-quantitative tissue distribution can be determined by whole body autoradiography [17] (see *Figure 3.10*).

(c) Formation of metabolites. Most investigations on the formation of metabolites from radiolabeled drugs involves extraction of the metabolites from a biological sample (usually plasma, urine or feces) followed by chromatographic analysis and identification. The extraction should be as mild as possible to avoid decomposition of any labile radiolabeled metabolites. The three most commonly used chromatographic systems are thin layer chromatography, high performance liquid chromatography (HPLC) and gas liquid chromatography (GLC). Radiolabeled metabolites on thin layer chromatography can be quantified by autoradiography while both HPLC and GLC can be linked to on-line counting. The identity of the radiolabeled metabolites can be determined by reference to known authentic standards, preferably on at least two chromatographic systems.

5.3.5 Endocytosis

Receptor mediated endocytosis involves the internalization of macromolecular ligands after binding to specific, cell surface receptors. The internalized ligand and receptor are then processed through a series of intracellular compartments (coated vesicles, early and late endosomes, etc.) prior to either degradation in the lysosomes or recycling back to the plasma membrane [18]. Thus, tracing receptor mediated endocytosis with radiolabeled ligands by pulse-chase techniques involves following the radiolabel through various subcellular organelles rather than through metabolic intermediates. These subcellular organelles can be separated by density gradient centrifugation.

In order to trace receptor mediated endocytosis, the macromolecule is usually radiolabeled with ^{125}I. A typical experiment to trace receptor mediated endocytosis in, for example, an isolated perfused rat liver, would involve pulsing the liver with a known amount of the ^{125}I-labeled ligand under single pass conditions [19]. First pass uptake of the ligand can be calculated by subtracting the amount of radiolabel recovered in the perfusate from that infused in the original pulse. At various times after the pulse of ^{125}I-labeled ligand, the liver is perfused with ice-cold medium in order to arrest further intracellular processing. The liver is then homogenized in a suitable ice-cold buffer and subcellular fractions isolated by density gradient centrigation (see [20]). After harvesting the gradient as multiple fractions, subcellular organelles are identified by reference to marker enzymes and ^{125}I associated with organelles determined by machine counting. Thus, the passage with time of ^{125}I-labeled ligands through different subcellular organelles can be determined.

If the internalized ^{125}I-labeled ligand is a relatively large protein of molecular weight >3–5 kd, intracellular degradation of the ligand can be assessed by determining the acid precipitability of the radiolabel. Intact ^{125}I-labeled protein will precipitate with trichloroacetic acid whilst small molecular weight ^{125}I-labeled products formed by proteolysis in the lysosomes will be acid-soluble.

5.4 Use of radioisotopes to determine *in vivo* kinetics

Radioactive tracers have been used to determine the kinetics of many processes *in vivo*. In this section we will consider three applications, namely: glucose kinetics; pharmacokinetics; and blood flow kinetics.

5.4.1 Glucose kinetics

The two standard methods of estimating rates of glucose utilization *in vivo* by compartmental analysis involve either a single bolus intravenous injection of radiolabeled glucose or a continuous intravenous infusion [21]. Blood samples are taken at intervals from the beginning of the experiment and the specific activity of glucose determined in d.p.m./μmol of glucose. The bolus injection considered here assumes that blood glucose concentrations are in steady state, that is, the rate of input equals the rate of output.

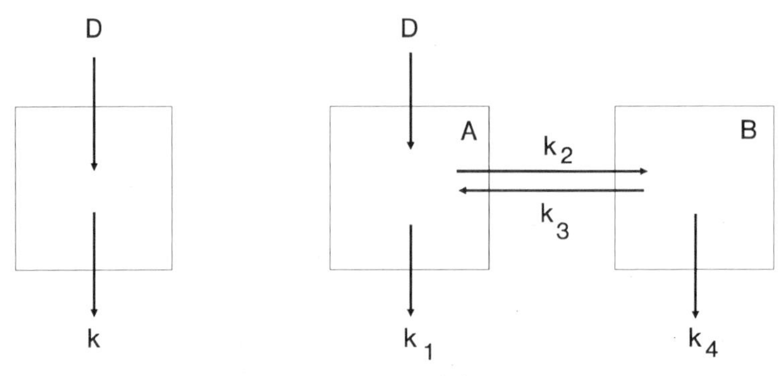

FIGURE 5.5: *(**a**) A dose of radiolabeled glucose, D, is administered into the pool and is irreversibly disposed with a rate constant k. (**b**) A dose of radiolabeled glucose, D, is administered into pool A and is irreversibly disposed with a rate constant k_1, or diffuses into pool B with a rate constant k_2. It then diffuses back to pool A or is irreversibly disposed from pool B with rate constants k_3 and k_4 respectively.*

If we consider a single pool system (*Figure 5.5a*) into which we administer a dose, D, of radiolabeled glucose, the specific activity, S, at time t is given by,

$$S = S_0 e^{-k \cdot t},$$

where S_0 is the specific activity at zero time and k is the rate constant for the irreversible disposal of glucose. This system would hold if glucose was entirely metabolized in the blood. However, glucose is transported into several other body pools and we must consider a more complex case. For example, if we consider a two pool system (*Figure 5.5b*), the specific activity at time t is given by,

$$S = A_1 e^{-k_1 \cdot t} + A_2 e^{-k_2 \cdot t} + A_3 e^{-k_3 \cdot t} + A_4 e^{-k_4 \cdot t},$$

where A_1, A_2, A_3 and A_4 are constants and their sum is S_0. In the case of glucose kinetics, even this model is too simple and we must consider a hypothetical model of n pools. Now,

$$S = \Sigma(A_i e^{-k_i \cdot t})$$

where A_i represents the constants A_1 to A_n and k_i represents the rate constants k_1 to k_n. The area under the specific activity–time curve is given by,

$$\int_0^\infty S dt = \int_0^\infty [\Sigma(A_i e^{-k_i \cdot t})] dt = \Sigma(A_i / k_i),$$

and the gradient of the specific activity–time curve at any time t is given by,

$$[\Sigma(A_i e^{-k_i \cdot t})] ds/dt = \Sigma(A_i k_i e^{-k_i \cdot t}) .$$

Thus, the gradient at zero time is $\Sigma(A_i k_i)$.

Using these expressions it is possible to calculate several parameters of glucose metabolism from the coefficients and exponents of the specific activity–time curve. These are:

(1) The rate of irreversible disposal of glucose. This is effectively equal to the rate of glucose phosphorylation by hexokinase and glucokinase, or, since steady state conditions are assumed, the replacement rate, that is, the rate of glucose-6-phosphatase.

(2) The mass of the sampling pool (i.e. the mass of glucose in the blood and those pools in rapid equilibrium with it).

(3) The rate of outflow from the sampling pool (by catabolism and exchange with the periphery).

(4) The total body mass of glucose.

(5) The minimum body mass and minimum transit time (i.e. the mass and transit time if it is assumed that catabolism of glucose takes place solely in the sampling pool).

(6) The maximum body mass and maximum transit time (i.e. the mass and transit time if it is assumed that catabolism of glucose occurs at a site or region removed as far as possible from the sampling pool).

These parameters can be calculated from the expressions given in *Table 5.1*.

Thus, a typical experiment involves taking blood samples (often via an indwelling cannula) after intravenous injection of a bolus of radiolabeled glucose. The specific activity of blood glucose is determined in d.p.m./µmol of glucose. This involves determining blood glucose concentrations by a standard enzymic assay and counting radioactivity associated specifically with blood glucose. Radiolabeled metabolites of glucose must be removed before counting. If ^{3}H-labeled glucose is used, freeze-drying of samples will remove ^3H_2O; if ^{14}C-labeled glucose is used, treatment with a mixed ion exchange resin will remove most ^{14}C-labeled metabolites (pyruvate, lactate, etc.). Specific activity–time curves (see *Figure 5.6a*) can then be fitted to exponential functions by iterative computer software.

The choice of the number of exponential terms in the mathematical function to fit the experimental data depends on the degree of experimental error [22]. A bi-exponential function is often found to give the best fit (see *Figure 5.6b*). The various kinetic parameters of glucose metabolism can then be calculated from the computed coefficients and exponents using the equations given in *Table 5.1*. Alternatively, the coefficients and exponents can be derived from a semi-logarithmic plot of the specific activity–time curve. Taking natural logarithms of the bi-exponential function,

$$S = A_1 e^{-k_1 t} + A_2 e^{-k_2 t} ,$$

yields,

$$\ln S = \ln A_1 - k_1 t + \ln A_2 - k_2 t .$$

A plot of ln S versus t will yield a curve which can be resolved into two straight lines (see *Figure 5.6b*); the intercepts on the ln S axis will give the coefficients (ln A_1 and ln A_2) and the gradients will give the exponents (k_1 and k_2).

TABLE 5.1: *Equations used for the calculation of the parameters of glucose metabolism*

Parameter	Symbol	Equation
Rate of irreversible disposal	R	$\frac{\text{dose}}{\text{area}} = \frac{D}{\Sigma(A_i/k_i)}$
Mass of sampling pool	M_s	$\frac{\text{dose}}{S_0} = \frac{D}{\Sigma(A_i)}$
Rate of outflow from sampling pool	R_s	$\frac{\text{dose} \times (\text{slope})_{t=0}}{[S_0]^2} = \frac{\Sigma(A_i k_i)D}{[\Sigma(A_i)]^2}$
Total body mass	M	$\frac{1}{k_n}R = \frac{D}{k_n \Sigma(A_i/k_i)}$
Minimum transit time	T_{min}	$\frac{\int_0^\infty S dt}{\text{area}} = \frac{(\Sigma A_i/k_i^2)}{\Sigma(A_i/k_i)}$
Minimum body mass	M_{min}	$T_{min} R = \frac{(\Sigma A_i/k_i^2)D}{[\Sigma(A_i/k_i)]^2}$
Maximum transit time	T_{max}	$\Sigma(1/k_i)$
Maximum body mass	M_{max}	$T_{max} \cdot R = \frac{\Sigma(1/k_i)D}{\Sigma(A_i/k_i)}$

Symbols are as follows: D, dose of radioactivity; S, specific radioactivity as a function of time; k, exponent of exponential terms (i.e. rate coefficients); k_n, smallest (final) exponent; A, coefficients of the exponential terms.

The choice of radiolabel for such experiments is critical. In order to determine accurately the kinetic parameters of glucose metabolism, an irreversible tracer must be used, that is, one which is not re-incorporated into glucose. Tritium is lost from glucose as ^3H_2O in the glycolytic pathway and, therefore, can be used as an irreversible tracer. However, ^3H from the 2 position is lost at the phosphoglucose isomerase reaction. Thus, [2-^{3}H]glucose will overestimate rates of irreversible disposal of glucose

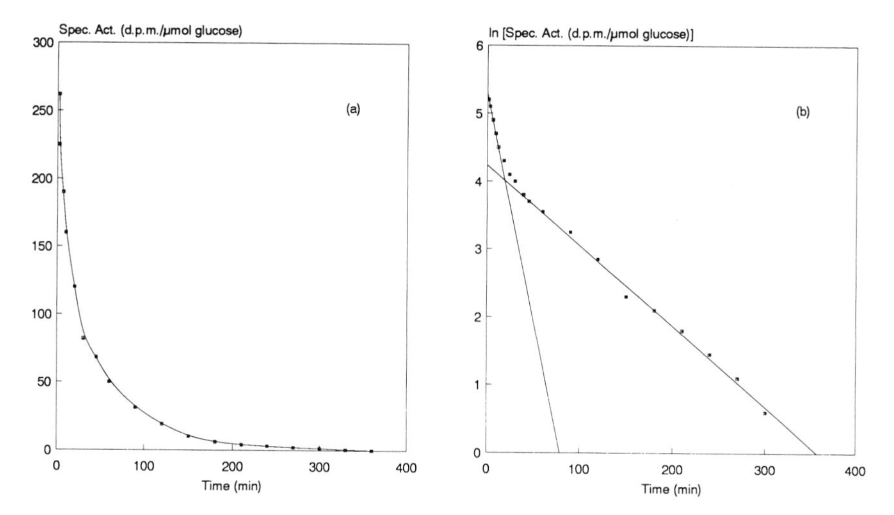

FIGURE 5.6: *(**a**) Disappearance of [2-^{3}H]glucose from the blood of rats after bolus intravenous administration. The equation of best fit is specific activity (at time t) = 200.e$^{-0.067t}$ + 70.e$^{-0.012t}$. (**b**) Data replotted on a semilogarithmic scale and resolved to produce two straight lines.*

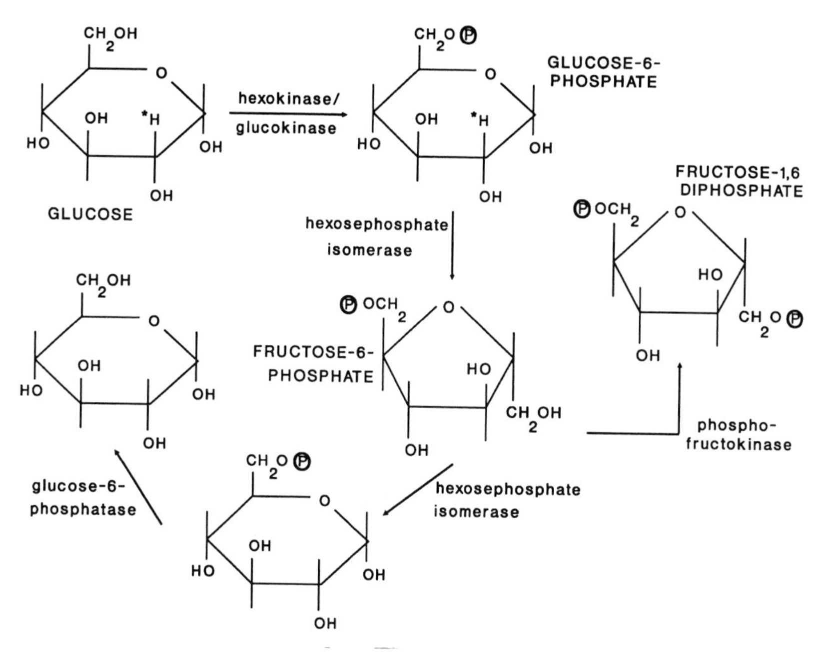

FIGURE 5.7: *Fate of [2-^{3}H]glucose in glycolysis. The ^{3}H atom is effectively lost at the hexosephosphate isomerase reaction by exchange with water; hexose phosphates recycling back to glucose via hepatic glucose-6-phosphatase are now unlabeled. Thus, [2-^{3}H]glucose will overestimate the rate of irreversible disposal of glucose (i.e. via phosphofructokinase) by the amount of ^{3}H lost in the substrate cycle glucose → glucose-6-phosphate → fructose-6-phosphate → glucose-6-phosphate → glucose.*

because ^{3}H will be lost in the substrate cycle glucose → glucose-6-phosphate → fructose-6-phosphate → glucose-6-phosphate → glucose (*Figure 5.7*). It is now generally agreed that [6-^{3}H]glucose is the most suitable irreversible tracer since ^{3}H is not lost until late in the glycolytic pathway [23]. On the other hand, if ^{14}C-labeled glucose is used, substantial amounts of ^{14}C-labeled pyruvate can be re-incorporated into glucose by hepatic gluconeogenesis. In this case, the apparent rate of irreversible disposal calculated from the specific activity–time curve for [U-^{14}C]glucose, will be the sum of the actual rate of irreversible disposal plus the rate of recycling of glucose carbon. This can be clearly seen from the specific activity–time curves in *Figure 5.8* which shows that [U-^{14}C]glucose disappears from plasma slower than [2-^{3}H]glucose after a bolus injection of [2-^{3}H, U-^{14}C]glucose.

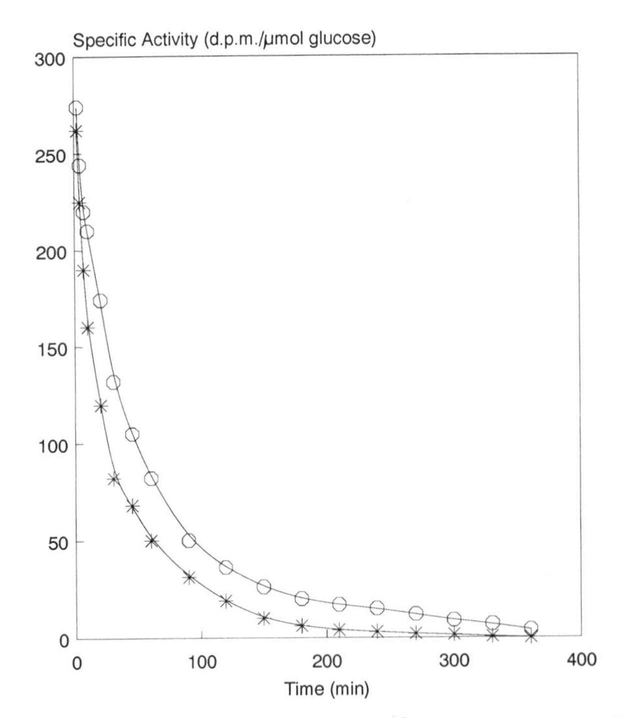

FIGURE 5.8: *Disappearance of [2-^{3}H]glucose (×) and [U-^{14}C]glucose (o) from the blood of rats after the intravenous administration of a bolus dose of [2-^{3}H, U-^{14}C]glucose.*

This can be used to advantage since the percentage recycling of glucose carbon is given by,

$$\frac{R_t - R_a \times 100}{R_t}$$

where R_t is the true rate of irreversible disposal calculated from the specific activity–time curve for an irreversible tracer (e.g. [6-^{3}H]glucose) and R_a is the apparent rate of irreversible disposal calculated from the specific activity–time curve for a reversible tracer (e.g. [U-^{14}C]glucose). Because the rate of irreversible disposal is inversely proportional to the area under the specific activity–time curve (see *Table 5.1*), the percentage recycling of glucose carbon can be calculated from the expression,

$$\frac{A_a - A_t}{A_a} \times 100$$

where A_a is the area under the specific activity–time curve for [U-^{14}C]glucose and A_t is the area under the corresponding curve for [6-^{3}H]glucose.

The percentage recycling of glucose is effectively a measure of hepatic gluconeogenesis. Such an approach has been used to show that recycling of glucose carbon (i.e. hepatic gluconeogenesis) is inhibited *in vivo* by a hypoglycemic agent [24]. Alternatively, gluconeogenesis *in vivo* can be estimated by determining the incorporation of radiolabeled substrates such as lactate into glucose [25].

5.4.2 Pharmacokinetics

Pharmacokinetics is the study of the rates of movement and biotransformation of a drug and its metabolites within the body. Radiolabeled drugs are often used to determine many kinetic parameters of drugs in experimental animals during preclinical drug development. However, measurement of total radioactivity in plasma or urine can easily produce confusing information. It is essential that the parent drug is first separated from radiolabeled metabolites by some sort of chromatographic system prior to machine counting.

Much of the theory of pharmacokinetics is the same as that already discussed for glucose kinetics, except that in this case we are interested in the behavior of an exogenous compound rather than an endogenous one. However, not all drugs are given as a single intravenous bolus injection; in fact most drugs are administered orally or they may be given by other routes such as subcutaneous or intramuscular injection. In such cases, a plot of plasma radioactivity associated with the parent drug versus time will show an absorptive phase and an elimination phase (*Figure 5.9*). Careful analysis of such curves can still yield much useful pharmacokinetic information.

We will not describe the derivation of various pharmacokinetic parameters; this is more than adequately covered in many textbooks on pharmacokinetics [see 26,27]. Suffice it to say that many pharmacokinetic parameters can be derived from plots of parent drug associated radioactivity in plasma versus time; for example, biological half-life (the time taken for the plasma concentration of a drug to decrease by half), volume of distribution (the size of a single distribution compartment of a drug) and clearance (the distribution volume totally cleared of drug per unit time). Similarly, if a drug is wholly or partly excreted in urine, much

FIGURE 5.9: A typical graph showing the variation of plasma radioactivity of a drug with time after a single oral dose. Note that the peak level represents the time at which the rate of absorption equals the rate of elimination.

useful information can be obtained from plots of parent drug associated radioactivity in urine versus time.

5.4.3 Blood flow kinetics

Many variations of the indicator dilution technique have been used to measure blood flow. The indicator dilution technique literally relies on administering a tracer (usually radiolabeled) into the circulation and measuring its dilution at a sampling site downstream from the site of administration. The method assumes that the tracer is inert, is well mixed in the blood, does not leave the circulation between the administration and sampling sites and does not disturb the hemodynamics.

Let us imagine blood flowing in a vessel at F ml/min. If a radiolabeled tracer is administered at a constant infusion rate of I ml/min and at an activity of D_{in} c.p.m./ml, then,

$$\text{input rate} = ID_{in} \text{ c.p.m./min.}$$

Blood samples are taken at a sampling site downstream from the infusion site until a constant, maximum steady state concentration of radioactivity is achieved (D_{out} c.p.m./ml). Now,

$$\text{output rate} = (F+I)D_{out} \text{ c.p.m./min.}$$

At steady state, the input rate must equal the output rate:

$$ID_{in} = (F+I)D_{out}.$$

If the infusion rate (I) is small compared to blood flow (F), then $(F+I) \sim F$, that is,

$$F = \frac{ID_{in}}{D_{out}} \text{ ml/min} \quad (1)$$

Thus, knowing the specific activity of the infused tracer (D_{in} c.p.m./ml), its infusion rate (I) and the specific activity at the downstream sampling site (D_{out} c.p.m./ml), the flow rate can be calculated.

This approach assumes that the tracer does not recirculate when its blood concentration would continually rise. In practice it is difficult to achieve steady state at the sampling site before recirculation occurs. An alternative approach is to administer the radiolabeled tracer as a bolus and to withdraw a blood sample from a downstream site at a constant, but small, rate; this is the so-called reference sample method. *Equation 1* can still be used to calculate blood flow but now D_{in} = the bolus dose of radiotracer (c.p.m.) and D_{out} = total recovery of radiotracer (c.p.m.) in a downstream sample withdrawn at I ml/min. In order for this method to be valid, the reference sample must be taken for a sufficient time to allow the 'tail' of the bolus of radiotracer to pass the sampling site before the 'head' of the bolus recirculates. Again, this is difficult to achieve in practice and complicated corrections have to be made for recirculation [28].

Carbonized microspheres are often used as a non-recirculating tracer. These are inert particles and when their diameter is >15 µm, they become trapped in arterioles on their first pass through tissues. Thus, the use of such tracers obviates any problems in blood flow measurements due to recirculation of the tracer. When used to study blood flow, microspheres are invariably radiolabeled with a γ-emitting isotope for ease of counting. Several isotopes have been used and include ^{51}Cr, ^{57}Co, ^{85}Sr, ^{99m}Tc, ^{125}I and ^{141}Ce. For example, ^{85}Sr-labeled microspheres have been used to determine cardiac output in rats [29]. The microspheres are given as a short (45 sec) bolus into the left ventricle via a cannula passed retrogradely from the right carotid artery. The reference blood sample is removed via a femoral artery cannula into a motor-driven syringe at a known rate. Starting a few seconds before the microsphere injection, the reference blood sample is withdrawn for 2 min until the entire bolus of microspheres has passed the sampling site. Knowing the total c.p.m. injected and the total c.p.m. recovered in the reference sample, cardiac output can now be calculated from *equation 1*.

Radiolabeled microspheres can also be used to determine regional blood flow in various organs. The animal is killed 2–3 min after intraventricular injection of microspheres, organs removed and trapped radioactivity determined [30]. Regional blood flow (ml/min) can then be calculated from the expression,

$$\frac{\text{organ radioactivity (c.p.m.)}}{\text{total radioactivity injected (c.p.m.)}} \times \text{cardiac output (ml/min).}$$

Another application of radiolabeled microspheres is to measure portal-systemic shunting in liver disease in experimental animals [31]. In liver cirrhosis, relatively large proportions (10–90%) of incoming blood from the hepatic portal vein is literally 'shunted' around the liver via collateral blood vessels. Given the central role of the liver in maintaining homeostasis, portal-systemic shunting is clearly of much significance and worthy of quantification. In this application, ^{57}Co-labeled microspheres are injected directly into the splenic pulp. In the absence of portal-systemic shunting, the microspheres will be trapped in the liver whilst significant shunting will lead to the microspheres becoming trapped in the next capilliary bed, that is, the lungs. Thus, if the animal is killed 10 min after the injection of microspheres and radioactivity determined in excized liver and lungs, portal-systemic shunting (as a percentage of portal blood flow) is given by the expression,

$$\frac{\text{lung radioactivity (c.p.m.)}}{\text{(lung + liver) radioactivity (c.p.m.)}} \times 100.$$

Recently, a method of quantifying portal-systemic shunting has been developed which obviates the need to remove organs for radioactivity counting [32]. The method involves the injection of ^{99m}Tc-labeled methylene diphosphonate into the splenic pulp. This is a small molecular weight tracer which will pass unhindered through the liver to the lungs. Thus, when detected by a sodium iodide scintillation detector positioned over the lungs, this will represent total flow (i.e. portal plus collateral flow). A second injection of ^{99m}Tc-labeled human albumin microspheres (diameter 20 µm) is then given. Detection of this radioactivity in the lungs represents shunted blood flow which can be expressed as a percentage of total flow. Since albumin microspheres are biodegradable, such a method offers the possibility of making serial measurements of portal-systemic shunting and may even be applicable to man.

5.5 Use of radioisotopes to study membrane transport processes

Membrane transport can be studied at all levels of biological organization ranging from the intact animal (or plant) to isolated cells, subcellular organelles and sealed membrane preparations, and even in model membrane preparations such as liposomes and black lipid films [33]. A common approach to studying transport across biological membranes is to use a radiolabeled permeant. The same considerations apply to the choice of radioisotope for tracing membrane transport apply to tracing metabolic processes.

5.5.1 Transport into intact tissues

^{3}H- or ^{14}C-labeled amino acids, monosaccharides, etc. have been extensively used in studying the intestinal uptake of nutrients. Experiments carried out *in vivo* usually involve perfusing the entire intestine (or defined segments) with the radiolabeled permeant in isotonic solution [34]. Measurement of radioactivity remaining in the efferent perfusate allows uptake to be determined. However, water will also be absorbed and this must be corrected for by including a non-absorbable marker in the perfusion fluid. If the permeant under investigation is ^{3}H-labeled, ^{14}C-labeled polyethylene glycol (PEG) 4000 can be included; both isotopes in the efferent perfusate can then be determined by dual label scintillation counting. Rates of uptake (μmol/min/cm of intestine) of the permeant can then be calculated from the expression,

$$\frac{V_i(^3H_i - {}^3H_r)\, {}^{14}C_i/{}^{14}C_r}{SAlt}$$

where, V_i = volume of test perfusate entering the intestine per collection period; ^3H_i = d.p.m./ml of ^{3}H-labeled permeant entering the intestine; ^3H_r = d.p.m./ml of ^{3}H-labeled permeant recovered from the intestine; $^{14}C_i$ = d.p.m./ml of [^{14}C]PEG 4000 entering the intestine; $^{14}C_r$ = d.p.m./ml of [^{14}C]PEG 4000 recovered from the intestine; SA = initial specific activity of ^{3}H-labeled permeant entering the intestine (d.p.m./μmol); l = length of intestine perfused (cm), and, t = time of collection period (min).

Quantitative autoradiography can also be used to localize transport processes, for example, the distribution of amino acid transport along the villous of the rat jejunum [35]. In this application, jejunum tissue is exposed to the ^{3}H-labeled amino acid for up to 45 sec and then fixed by incubation in 4% (w/v) glutaraldehyde in phosphate-buffered saline, pH 7.3, for 1 h. After washing in phosphate-buffered saline to remove excess fixative, the tissue is embedded in a suitable resin and sections (10 μm) cut. The sections are coated in a nuclear emulsion (e.g. Kodak NTB-2) and left for 14–21 days to develop. The tissue can then be examined by microscopy when amino acid uptake will appear as densely stained areas. Uptake along the length of the villous can be quantified by scanning microdensitometry.

5.5.2 Transport into membrane vesicles

An alternative approach to studying intestinal transport is to use a sealed vesicle preparation from small intestine brush border membranes [36]. The radiolabeled permeant is incubated with a suspension of the vesicles in buffer. Uptake can be stopped by the addition of ice-cold buffer and radioactivity taken up into the vesicles can be separated from that in the medium by vacuum filtration through filter discs. After washing the discs in ice-cold buffer, radiolabeled permeant taken up into the vesicles can be determined by counting the discs. When measuring rates of uptake, it is imperative that the rate remains constant throughout the incubation period and this should be validated. Passive and facilitated transport processes rapidly achieve equilibrium, when the radiolabeled permeant diffuses out of the vesicle at a rate equal to its rate of uptake. Thus, very short incubation times (of the order of seconds) are often required.

Liposomes have been used extensively to investigate the transport properties of the ion-translocating antibiotics (e.g. valinomycin, gramicidin A). In such cases, liposomes are usually prepared containing a radiolabeled permeant and efflux of the permeant is monitored. For example, valinomycin has been shown to act as a mobile carrier in stimulating $^{42}K^{+}$ efflux from preloaded liposomes [37].

5.5.3 Transport into cells

A similar approach can be used to study uptake into cells. Whilst it is possible to use vacuum filtration to separate radiolabeled permeant taken up into cells from that in the medium, centrifugation is often used because cells sediment much more rapidly than vesicles. Centrifugation

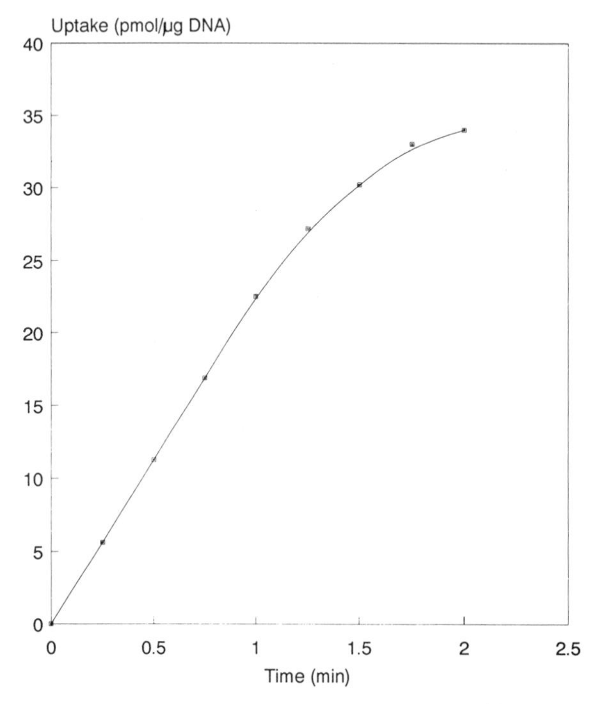

FIGURE 5.10: Time course of uptake of [24-^{14}C]taurocholate into hepatocytes cultured on Thermanox cover slips. Rates of uptake can be determined as the gradient of the line up to 1 min. After this time, significant release of bile salt back into the medium occurs and uptake becomes non-linear.

through a cushion of silicone oil achieves almost total separation of radiolabeled permeant taken up from that in the medium. Alternatively, if cultured cells are used and these are attached to a plastic tissue culture dish or a cover slip, simply removing the medium and washing the cells in ice-cold buffer before counting allows the amount taken up to be determined. An extra consideration when using cells is that some cell types are polar, that is, permeants are taken up at one side of the cell and secreted from the cell at the opposite side. For example, bile salts are actively transported into hepatocytes across the sinusoidal membrane (that part of the plasma membrane facing the blood), and are then actively transported out of the cell into the forming bile across the canalicular membrane. Thus, when determining rates of uptake of radiolabeled bile salts into isolated hepatocytes, short incubation times are required (see *Figure 5.10*).

References

1. Wood, S.G., John, B.A., Chausseaud, L.F., Johnstone, I., Biggs, S.R., Hawkins, D.R., Priego, J.G., Darragh, A. and Lambe, R.L. (1983) *Xenobiotica,* **13**, 731–741.
2. Belleau, B. (1965) in *Isotopes in Experimental Pharmacology* (L.J. Roth, ed.). University Press, Chicago, p. 469–481.
3. Mazelis, M. (1971) *Methods Enzymol.,* **XVIIB**, 606–608.
4. Axelrod, J. (1971) *Methods Enzymol.,* **XVIIB**, 761–769.
5. Fleischer, B., Fleischer, S. and Ozawa, H. (1969) *J. Cell Biol.,* **43**, 59–79.
6. Philipp, B.W. and Shapiro, D.J. (1979) *J. Lipid Res.,* **20**, 588–593.
7. Princen, H.M.G., Meijer, P., Kwekkeboom, J. and Kempen, H.J.M. (1988) *Anal. Biochem.,* **171**, 158–165.
8. Billheimer, J.T., Tavani, J.T. and Hes, W.R. (1981) *Anal. Biochem.,* **111**, 331–335.
9. Garrison, J.C. (1983) *Methods Enzymol.,* **99**, 20–36.
10. Yin Jen Wang, J. and Baltimore, D. (1983) *Methods Enzymol.,* **99**, 373–378.
11. Auruch, J. and Wallach, D.F.H. (1971) *Biochim. Biophys. Acta,* **233**, 334–347.
12. Krebs, H.A. (1970) *Perspect. Biol. Med.,* **14**, 154–170.
13. Bremer, J. and Davis, J.E. (1978) *Biochim. Biophys. Acta,* **528**, 269–275.
14. Senior, A.E., Holland, P.C. and Sherratt, H.S.A. (1975) in *A Symposium on Hypoglycin* (E.A. Kean, ed.). Academic Press, New York, p. 109–119.
15. Claus, T.H., Pilkis, S.J. and Park, C.R. (1975) *Biochim. Biophys. Acta,* **404**, 110–123.
16. Hawkins, D.R. (1988) in *Radiochemicals in Biomedical Research* (E.A. Evans and K.G. Oldham, eds). Wiley, Chichester, p. 14–55.
17. Curtis, C.G., Cross, S.A.M., McCulloch, R.J. and Powell, G.M. (1981) *Whole Body Autoradiography.* Academic Press, London.

18. Wileman, T., Harding, C. and Stahl, P. (1985) *Biochem. J.*, **232**, 1–14.

19. Perez, J.H., Branch, W.J., Smith, L., Mullock, B.M. and Luzio, J.P. (1988) *Biochem. J.*, **251**, 763–770.

20. Ford, T.C. and Graham, J.M. (1991) *An Introduction to Centrifugation.* BIOS Scientific Publishers, Oxford.

21. Katz, J., Rostami, H. and Dunn, A. (1974) *Biochem. J.*, **142**, 161–170.

22. Shipley, R.A. and Clark, R.E. (1972) *Tracer Methods for in vivo Kinetics: Theory and Applications.* Academic Press, London.

23. Katz, J., Golden, S., Dunn, A. and Chenoweth, M. (1976) *Hoppe-Seyler's Z. Physiol.* **357**, 1387–1394.

24. Osmundsen, H., Billington, D., Taylor, J.R. and Sherratt, H.S.A. (1978) *Biochem. J.*, **170**, 337–342.

25. Radzuik, J. (1982) *Federation Proc.*, **41**, 88–90.

26. Clark, B. and Smith, D.A. (1986) *An Introduction to Pharmacokinetics.* Blackwell Scientific Publications, Oxford.

27. Bourne, D.W.A., Triggs, E.J. and Eadie, M.J. (1986) *Pharmacokinetics for the Non-Mathematical.* M.T.P. Press, Lancaster.

28. Griffiths, R. (1990) in *Radioisotopes in Biology: A Practical Approach* (R.J. Slater, ed.). IRL Press, Oxford, p. 109–136.

29. Malik, A.B., Kaplan, J.E. and Suba, T.M. (1976) *J. Appl. Physiol.*, **40**, 472–475.

30. Lee, S.S., Girod, C., Valla, D., Geoffroy, P. and Lebrec, D. (1985) *Am. J. Physiol.*, **249**, G528–G532.

31. Chojker, M. and Groszmann, R.J. (1981) *Am. J. Physiol.*, **240**, G371–G375.

32. Yates, J., Nott, D.M., Billington, D., Shields, R. and Jenkins, S.A. (1991) *Gut*, **32**, A553.

33. Stein, W.D. (1990) *Channels, Carriers and Pumps: An Introduction to Membrane Transport.* Academic Press, New York.

34. Martines, D., Morris, A.I. and Billington, D. (1989) *Alcohol Alcoholism*, **24**, 525–531.

35. Cheeseman, C.I. (1986) *Am. J. Physiol.*, **251**, G636–G641.

36. O'Neill, B., Weber, F., Hornig, D. and Semenza, G. (1986) *FEBS Lett.*, **194**, 183–188.

37. Ovchinnikov, Y.A. (1979) *Eur. J. Biochem.*, **94**, 321–336.

6 Use of Radioisotopes in Molecular Biology

Radioisotopes are used extensively in molecular biology and autoradiography (see Section 3.3) is often used to either localize radioactivity in biological samples or to quantify radioactivity in samples where machine counting is impractical. This chapter is devoted to some of the applications of radioisotopes in molecular biology. Later books in the series will be concerned specifically with various aspects of molecular biology.

6.1 DNA sequencing

6.1.1 Background

From a knowledge of the base sequence of DNA in the gene, it is a simple matter to determine the amino acid sequence of the protein specified by the gene. In fact, nowadays it is quicker to determine the amino acid sequence of a protein from the sequence of bases in its gene rather than by direct amino acid sequencing. Thus, DNA sequencing has many applications and has been used, for example, to investigate the molecular basis of many inherited genetic disorders when a protein is either not expressed or is expressed in a non-functioning form.

Until recently all DNA sequencing employed radiochemical methods with autoradiographic detection. In practice, it is only possible to sequence DNA fragments of up to a maximum of 500 bases in length. Such fragments are produced by cleaving DNA with restriction enzymes. These enzymes cleave both strands of the double helix at specific sites and over 90 such enzymes have now been purified and characterized [1]. After sequencing the restriction fragments, the full DNA sequence can be built up by overlapping consensus sequences at different sites of cleavage. Two basic methods are available for sequencing restriction fragments and are now discussed.

6.1.2 Interruption of replication (Sanger dideoxy method)

This method utilizes 2',3'-dideoxynucleotide triphosphates (ddNTPs); these are specific terminators of DNA chain elongation [2]. ddNTPs can be incorporated normally into a growing DNA chain through their 5'-triphosphate but cannot form a phosphodiester bond with another deoxynucleotide triphosphate (dNTP) because the 3'-hydroxyl is absent.

The dideoxy method relies upon copying (replicating) the single-stranded DNA to be sequenced. Thus, the single-stranded DNA is incubated with a short oligonucleotide primer of known sequence, the Klenow fragment of DNA polymerase 1, all four dNTPs and a small amount of a ddNTP. If the correct ratio of ddNTP:dNTP is used and, for example, ddATP is used, the products will be a series of chains each terminating at A. In practice, four separate incubations are needed, each containing a different ddNTP. These will produce four series of chains terminating at each base. The resultant fragments can be separated according to their molecular weight on four lanes of a polyacrylamide gel [3,4].

If one or more of the dNTPs are radiolabeled, the fragments can be visualized on gels by autoradiography. dNTPs radiolabeled in the α-phosphate with ^{32}P (see *Figure 6.1a*) are commercially available. It is essential that ^{32}P is in the α-position since this is the phosphate group which is incorporated into the DNA backbone. Alternatively, phosphorothioate analogs radiolabeled with ^{35}S can be used (see *Figure 6.1b*); these analogs are incorporated by DNA polymerase 1 with high efficiency. Labeling with ^{35}S gives better resolution on autoradiography

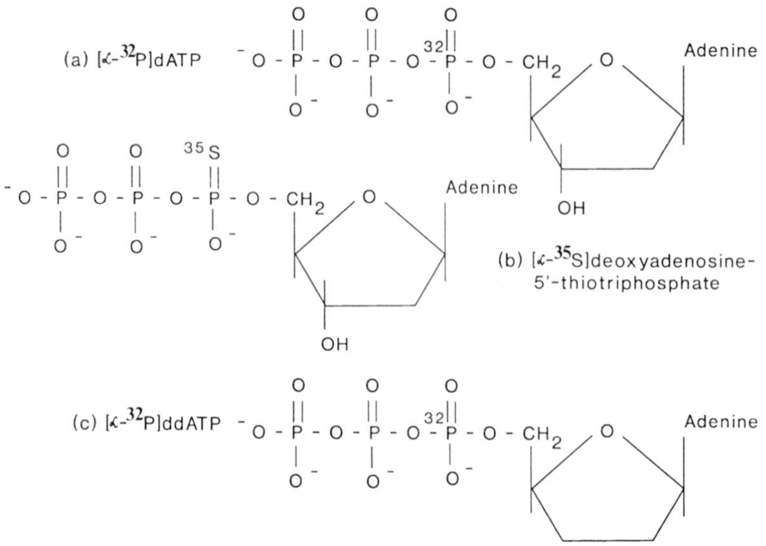

FIGURE 6.1: *Structural formulae of some radiolabeled nucleotides used in DNA sequencing.*

than with ^{32}P such that longer sequences can be read from each autoradiogram. However, ^{35}S requires longer exposure times than ^{32}P (see Section 3.3). Another approach is to use ddNTPs radiolabeled in the α-phosphate. However, because each fragment will only be radiolabeled at the terminal base, a high energy β-emitting isotope is required. Thus, ^{32}P-labeled ddNTPs are used (see *Figure 6.1c*).

The shortest DNA fragment moves the greatest distance on polyacrylamide gel electrophoresis. Thus, knowing the terminal base of each fragment from which ddNTP was included, and knowing that A base-pairs with T and G base-pairs with C, the sequence of the original DNA strand can now be read upwards from the bottom of the gel (see *Figure 6.2*).

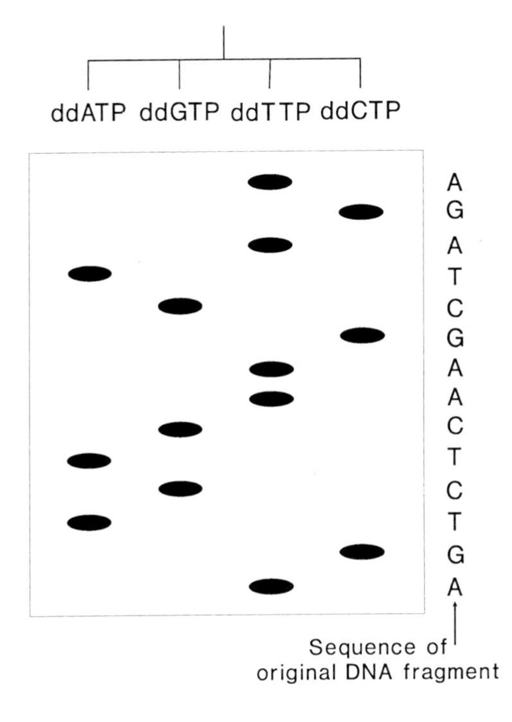

FIGURE 6.2: *DNA sequencing by the Sanger dideoxy method. The schematic autoradiogram shows the radiolabeled fragments produced by incubation with ddNTP chain terminators. The sequence of bases in the original DNA fragment is read upwards by base-pairing to the terminating dideoxy residue.*

Recently, a variant of the dideoxy method has been introduced which does not utilize radioisotopes; instead, a fluorescent label is attached to the chain terminating ddNTPs [5]. If a different colored fluoresence emitter is used for each of the four ddNTPs, only one incubation is required. This will consist of the DNA fragment to be sequenced, an oligonucleotide primer, the Klenow fragment of DNA polymerase 1, all four dNTPs and all four fluorescently-labeled ddNTPs. When the resultant fragments are electrophoresed on a polyacrylamide gel, the separated bands of DNA pass the detection system (a laser beam) targeted at a specific area of the gel. As each band passes the detection system, the sequence can now be read as the base pair to the fluorescently-labeled ddNTP terminator.

6.1.3 Chemical degradation (Maxam and Gilbert method)

In the chemical degradation method of sequencing restriction fragments, DNA is end-labeled using polynucleotide kinase at the 5'-hydroxyl terminus (see Section 2.4.4). Since only one radioactive atom is incorporated per restriction fragment, ^{32}P is usually used because it is a high energy β-emitting isotope. The single-stranded DNA fragment is then subjected to a series of reactions which modifies one or two of the bases and results in strand breaks at specific sites [6]. The conditions are controlled such that only a few sites are broken in any one DNA strand. Thus, a series of ^{32}P-labeled fragments are produced extending from the ^{32}P-label on the 5'-terminal base to the site of cleavage.

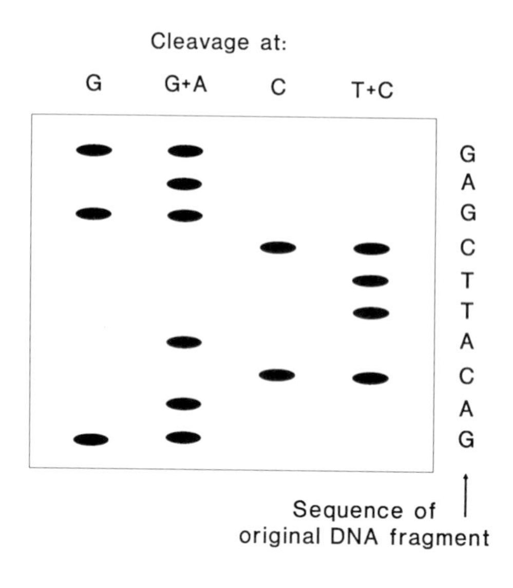

FIGURE 6.3: DNA sequencing by the Maxam and Gilbert chemical degradation method. The schematic autoradiogram shows the radiolabeled fragments produced by chemical cleavage at either A, G+A, C and T+C of [5'-^{32}P]end-labeled DNA fragments. The sequence of bases in the original DNA fragment is directly read upwards from the bottom of the gel.

Four series of reactions are required which cleave at different bases. A commonly used variant of the Maxam and Gilbert method is to use reactions which cleave at G, A plus G, C, and T plus C. When the resultant fragments are resolved on polyacrylamide gel electrophoresis and visualized by autoradiography, the base sequence can now be read upwards from the bottom of the gel (see *Figure 6.3*). Note that this method allows the base sequence to be read directly, whilst in the dideoxy method, the sequence is read as the base pair to the terminating nucleotide.

6.2 Unscheduled DNA synthesis

DNA contains many potentially reactive sites and its structure can be deleteriously modified in many ways (e.g. by reaction with chemical mutagens or free radicals, or by physical agents such as u.v. light and α-, β- and γ-radiation). All organisms maintain their genetic integrity by repairing DNA damage. However, some DNA-repair mechanisms are error prone and cells may survive potentially lethal DNA damage at the cost of introducing a mutation. Induced DNA repair can be used as a means of detecting the ability of chemicals and physical agents to reach and damage DNA, and is effectively an assay of genotoxicity. Unscheduled DNA synthesis (UDS) is the term used to describe this repair process; the term derives from the fact that cells are not undergoing cell division and yet are synthesizing DNA, that is, the synthesis is 'unscheduled' compared with the much greater 'scheduled' DNA synthesis which preceeds mitosis ('S' phase in the cell cycle).

UDS involves a damage-specific endonuclease which recognizes the distortion and excizes the damaged sequence of nucleotides such that the DNA double helix now contains a gap in one strand. This gap is filled by polymerases which base pair to the undamaged complete strand. The repair is completed by ligases which seal the new nucleotide sequence into the incomplete strand. UDS in cell cultures can be quantified by incubating the cells with a radiolabeled nucleotide; [^{3}H]thymidine is used because this nucleotide does not occur in RNA. The amount of radiolabel incorporated into nuclear DNA is then measured. This is usually done by autoradiography such that cells in 'S' phase (which show massive incorporation of radiolabel due to 'scheduled' DNA synthesis) can be excluded.

UDS in primary cultures of rat hepatocytes is probably the most widely applied technique for assessing the potential genotoxicity of chemical and physical agents [7,8]. UDS assays have also been developed for other cell types, for example, pancreas [9], kidney [10], stomach [11] and various transformed cell lines. In its simplest form, a UDS assay involves exposing primary cell cultures to various concentrations of potential genotoxins dissolved in the culture medium. However, this method does not reflect the complex patterns of absorption, distribution, metabolism and excretion of compounds that occurs in whole animals. Some UDS assays using cell types other than hepatocytes incorporate a hepatic microsomal fraction into the culture medium to check for hepatic activation of possible genotoxins [12]. Also, factors such as chronic exposure, sex differences and different routes of exposure cannot be studied in this system. Thus, a potentially more valuable method of assessment of genotoxicity is to measure UDS in primary cell cultures derived from animals pretreated *in vivo* with suspect agents. This approach is now gaining favor for the routine assessment of the potential genotoxicity of chemical agents [13].

When using autoradiography to detect UDS, cells are invariably cultured on cover slips in suitable well plates. After cell attachment, the culture medium is replaced by one containing 0.2–0.4 MBq/ml of [^{3}H]thymidine and the cells are cultured for a further 1–24 h. Each coverslip is then

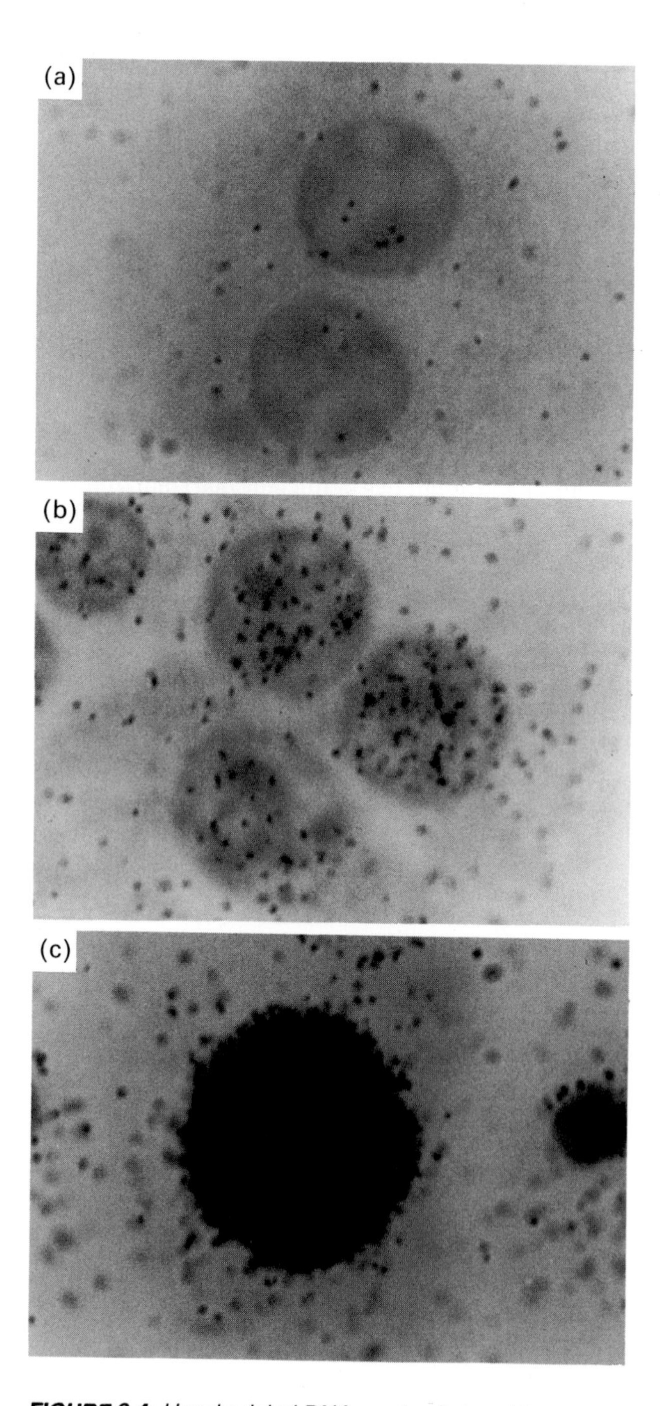

FIGURE 6.4: *Unscheduled DNA synthesis in rat hepatocytes. Cells were incubated with [^{3}H]thymidine and developed for autoradiographic analysis. (**a**) Control cells showing some nuclear grains and background cytoplasmic grains. (**b**) Cells from a rat pretreated with a genotoxic agent showing increased density of nuclear grains. (**c**) Cell in S-phase showing uncountable dense nuclear grains. Photographs by kind permission of Dr J.C. Kennelly.*

removed from its well, washed free of extracellular [^{3}H]thymidine in phosphate-buffered saline and incubated in 1% (w/v) trisodium citrate for 10 min to swell the nuclei to allow better quantification of nuclear grains. The cells are fixed on the cover slip by immersion in ethanol : acetic acid (3 : 1, v/v) and mounted cell surface up on a glass slide. The slides are then coated under a suitable safe light with a nuclear emulsion (e.g. Kodak NTB-2) and exposed for the recommended period (7 days) in a light-proof container at 4°C. The slides are developed in a dark room by immersion in Kodak D19 developer for 4 min, quickly washed in tap water and then fixed by immersion in Kodak F24 fixer. The resulting clear slides are rinsed in water and air-dried. The resultant nuclear grains, produced by the incorporation of [^{3}H]thymidine into repaired DNA, are counterstained with, for example, Harris' Alum haematoxylin, Pronin Y or Giemsa stain.

The slides can now be analyzed by eye using light microscopy or by connecting the microscope to an electronic counter. Grain counts are scored separately over the nuclear and cytoplasmic domains. Cells in S-phase engaged in replicative DNA synthesis are characterized by uncountable dense nuclear grains and are excluded from the scoring. Results are conventionally expressed as either grains per nucleus or grains per unit area of nucleus after correction for background cytoplasmic grains; 40–100 cells per slide are usually counted and average grain counts recorded. A positive genotoxic agent will produce an increased density of nuclear grains compared to corresponding controls. *Figure 6.4* demonstrates the detection of UDS in primary cultures of rat hepatocytes by autoradiography.

Although it is preferable to quantify UDS by autoradiography, it is possible to use liquid scintillation counting. In such cases, it is imperative to eliminate replicative DNA synthesis. This can be achieved by culturing cells to confluence such that cell division is considerably reduced by contact inhibition and by adding 10 mM hydroxyurea 1 h prior to the chemical exposure and the addition of [^{3}H]thymidine [12]. After exposure, cellular DNA is extracted, quantified and [^{3}H]thymidine incorporation determined by counting; results are expressed as d.p.m. per μg of DNA.

6.3 Assay of DNA damage and repair by nick translation

Many DNA-damaging carcinogens test negative in the unscheduled DNA synthesis assay. In addition, the UDS assay does not allow the type of damage or repair to be determined and, indeed, is likely to miss carcinogens which induce DNA damage that requires little or no strand resynthesis during repair. It is for these reasons that the nick translation assay was designed [14]. This assay detects both DNA damage (strand breaks) and repair in cultured cells and largely overcomes the problems of the UDS assay.

The principle of the nick translation assay is that DNA polymerase 1 binds to sites of DNA damage (nicks) containing free 3'-hydroxyl termini. Owing to the concerted action of the polymerase and the 5' to 3' exonuclease activity within the enzyme molecule, the polymerase moves (translates) along the helix incorporating bases from cellular deoxynucleotide triphosphates (dNTPs). The process is terminated at either a second strand interruption or a topological constraint, but not until long stretches of DNA have been so processed. If one of the cellular dNTPs is radiolabeled, incorporation into DNA can be quantified, effectively providing an assay of nick translation.

Initial reports on the applications of nick translation assays used cultured human fibroblasts or mouse mammary epithelial cells in free suspension [15,16]. Cell cultures are exposed to the potential carcinogen for an appropriate time and then harvested from culture flasks by trypsin treatment. The assay involves incubating permeabilized fibroblasts for 30 min in a medium containing *E. coli* DNA polymerase 1 (40 units/ml), 50 μM each of dATP, dGTP and dTTP and 0.2 MBq/ml of [^{3}H]dCTP. The cells must be permeabilized with, for example, lysophosphatidyl choline, in order to allow the DNA polymerase 1 and dNTPs to enter. However, excessive permeabilization can lead to high background levels of DNA damage in the absence of carcinogens. Exogenous DNA polymerase 1 is necessary to allow greater incorporation of dNTPs than would be afforded by cellular polymerases. At the end of the incubation, the assay mixture is filtered through filter discs followed by extensive washing with 5% (w/v) trichloroacetic acid. Acid-insoluble radioactivity (i.e. [^{3}H]dCTP incorporated into cellular DNA) is trapped on the filter discs and can be quantified by liquid scintillation counting. Unlike UDS assays, cells in S-phase do not affect the incorporation of [^{3}H]dCTP and thus addition of genotoxic agents such as hydroxyurea is unnecessary.

Results are best expressed as d.p.m. of [^{3}H]dCTP incorporated per 10^{5} cells. If cultures are seeded at the same initial cell density and passage number and grown to confluence, it is possible to express the results simply as dp.m./culture. An increased incorporation of [^{3}H]dCTP above corresponding controls is considered an indication of DNA strand breaks and repair.

More recent reports utilizing the nick translation assay have used autoradiography as the detection system [17,18]. In such cases, cells are cultured on cover slips, exposed to the potential carcinogen for an appropriate time and then fixed *in situ* with ethanol. The cover slips are then incubated for up to 1 h in a nick translation medium containing buffer, *E. coli* DNA polymerase 1 and all four dNTPs, one of which must be radiolabeled. After incubation, the cover slips are extensively washed and then developed for autoradiography as described for UDS (see

Section 6.2). Results are expressed as grains per nucleus or per unit area of nucleus, and increased grain counts are indicative of DNA strand breaks and repair.

One disadvantage of the nick translation assay is that possible carcinogens are added directly to culture medium. Thus, it is difficult to assess chronic exposure to potential carcinogens, or their possible conversion to active metabolites by other tissues. However, in a recent study, mouse sarcoma 180 cells implanted into the peritoneal cavity have been used in a nick translation assay [18]. The test compound can then be administered *in vivo*, and, after a suitable exposure period, the tumor cells are aspirated from the peritoneal cavity, smeared on to glass slides and processed for nick translation as before. This approach provides an *in vivo* nick translation assay analogous to the *in vivo* UDS assay.

6.4 *In situ* hybridization

The term *in situ* hybridization refers to the detection of specific nucleic acid sequences within tissues, cells or on chromosomes using radiolabeled sequence-specific DNA or RNA probes. Among the uses of this technique are the detection of specific gene sequences and their chromosome location [19] and the detection of viral DNA and their RNA transcripts [20]. *In situ* hybidization is a rather complicated technique and has been recently reviewed [21]; thus, only an overview will be presented here. The incorporation of radioisotopes into DNA and RNA has been described in Section 2.4.

Cells can either be cytocentrifuged onto poly-L-lysine hydrobromide coated glass slides or cultured directly on cover slips and then mounted on to slides. Cryostat sections (10 μm) of frozen tissue are mounted directly on to coated slides. The tissue/cells are fixed in an appropriate fixative prior to hybridization. It is now widely believed that single-stranded RNA probes (of between 50 and 250 bases in length) are more sensitive hybridization probes than double-stranded DNA probes. The choice of radioisotope for the probe is often a compromise between resolution and sensitivity for subsequent autoradiography. Thus, low energy β-emitting ^{3}H-labeled probes give the best autoradiographic resolution but often require exposure times of weeks, whilst high energy β-emitting ^{32}P-labeled probes give decreased autoradiographic resolution but require much shorter exposure times. ^{35}S-labeled probes represent a compromise between these two extremes.

Hybridization of fixed tissue/cells is carried out by coating slides with a small volume (10–20 μl) of the radiolabeled probe in buffer and incubating at 50–150°C in a humid atmosphere for upwards of 2 h. Non-hybridized radiolabeled probe is removed by several thorough washing cycles in buffer. Finally, the tissue/cells are dehydrated in increasing concentrations of ethanol prior to autoradiography.

The preparation of slides for autoradiography and subsequent staining is essentially similar to that for unscheduled DNA synthesis (see Section 6.2). The resultant grains, due to hybridization of the radiolabeled DNA or RNA probes into the cellular DNA or RNA, can now be viewed by microscopy. One of the advantages of using radioisotopes in this application is that results can be quantified by grain counting over particular cells/ nuclei.

6.5 Northern, Western and Southern blots

In order to determine the presence and quantity of specific macromolecules (protein, DNA or RNA) in a tissue or cell homogenate or in a subcellular fraction, the preparation is separated first according to molecular weight by electrophoresis on a gel [3,4]. Polyacrylamide gels are used for proteins whilst agarose gels are used for DNA and RNA. The separated components are then transferred (blotted) to a membrane and the specific macromolecule is determined and quantified on the membrane using a radiolabeled probe and autoradiography. Whilst blotting increases the complexity of the electrophoretic technique, it has substantial advantages. For example, membrane transfer avoids non-specific reactions of the radiolabeled probe with the gel matrix and, when bound to the surface of a thin membrane, the macromolecule is more accessible to the radiolabeled probe than if buried within the gel matrix.

The concept of blotting was first introduced by Southern [22] who blotted DNA from agarose gels on to nitrocellulose membranes by capillary transfer; the technique subsequently became known as 'Southern' blotting. The gel is placed on to a pad of filter papers moistened with buffer. The nitrocellulose membrane is then placed on top of the gel, taking care to exclude air bubbles which would interfere with transfer. A layer of dry absorbent paper is placed on top of the membrane and held in place with a weight. Buffer flows from the wet filter papers to the dry absorbent paper carrying the macromolecules from the gel to the membrane. When diazobenzyloxymethyl membranes were introduced, the technique became known as 'Northern' blotting and still used capillary transfer. Subsequently electroblotting was introduced where transfer of macromolecules (particularly proteins) is effected by electric current; this is colloquially known as 'Western' blotting. A nitrocellulose membrane is layered on to the gel (taking care to exclude air bubbles) and sandwiched between two layers of foam sponge material or filter papers and mounted on a stiff plastic grid. The entire sandwich is submerged in buffer in a tank and transfer is effected by applying a voltage gradient of 5–10 V/cm from electrodes in the tank. For 1.5 mm thick polyacrylamide gels, >90% of all proteins of molecular weights up to 10^5 are transferred within 16 h. The terms Southern, Northern and Western blotting are used nowadays to refer to the detection of DNA, RNA and proteins respectively.

The mechanism by which transferred polypeptide, DNA and RNA are bound to nitrocellulose membranes is poorly understood and, indeed, some components only bind with weak affinity and may be lost during subsequent processing of the membrane. This can be prevented by 'fixing' the membrane. This is achieved by immersion in, for example, Tris-buffered saline containing 1% (w/v) gelatine for 1 h or phosphate-buffered saline containing 0.5% (w/v) glutaraldehyde for 15 min. Diazobenzyloxymethyl membranes overcome this problem; negatively charged groups bind electrostatically to positively charged diazonium groups followed by a slow reaction via azo derivatives leading to covalent bond formation.

Antibody probes (preferably monoclonal) are used to identify specific polypeptides on membrane blots and are almost always radiolabeled with ^{125}I. Non-specific binding sites on the membrane are blocked by incubation with excess protein (e.g. bovine serum albumin) and the membrane is then incubated for several hours with the antibody. Whilst it is possible to ^{125}I-label this antibody, the most commonly used detection system in Western blotting is a double antibody system analogous to that used in immunoradiometric assays (see Section 4.1.9) [21]. In this case, the membrane is thoroughly washed after incubation with the unlabeled primary antibody, and the primary antibody bound to a specific polypetide on the membrane is then detected by incubation with a ^{125}I-labeled second

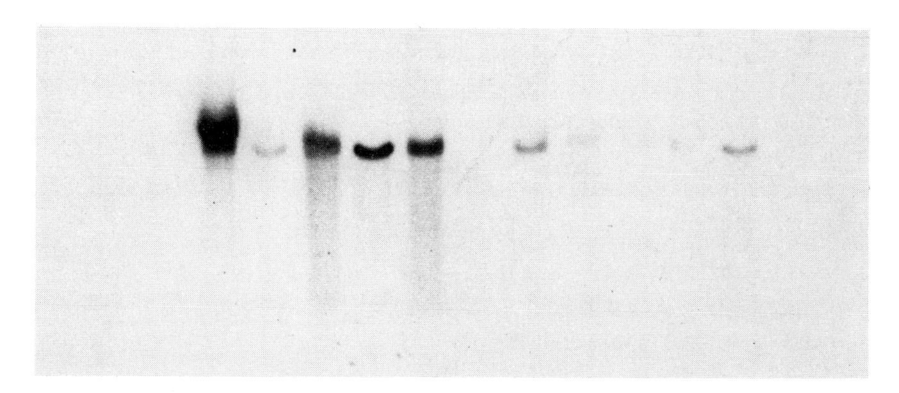

FIGURE 6.5: *A typical Northern blot analysis. Total RNA was extracted from several samples of human breast tumor tissue, electrophoresed on agarose and blotted onto a nylon support (Hyband-N). Hybridization of the nylon support to a ^{32}P-labeled cDNA clone to a specific estrogen-regulated mRNA species was carried out in the presence of 50% (w/v) formamide and 50 µg/ml salmon sperm DNA. After extensive washing, the filter was exposed to preflashed film for 2 days at –70°C. Photograph by kind permission of Dr P.A. Burke.*

antibody (anti-immunoglobulins) at approximately 2.4 MBq/ml. Controls should be run omitting the primary antibody. The ^{125}I-labeled antibody is then determined by direct autoradiography (see Section 3.3) of the membrane and may be quantified by densitometric scanning of the X-ray film. Non-radioactive labeled probes (e.g. fluoresence) are gaining in popularity.

Specific nucleotide sequences in DNA and RNA on membrane blots are detected by hybridization with sequence-specific, radiolabeled probes [21]. Membranes are first immersed for at least 4 h in buffer at 37°C containing denatured DNA from either herring or salmon sperm or from *E. coli* in order to saturate non-specific binding sites. The radiolabeled hybridization probe is then added and the incubation continued for up to 48 h. After thorough washing, the membrane is dried and subjected to autoradiography. As with *in situ* hybridization (see Section 6.4), the choice of radiolabel for the probe is invariably a compromise between resolution and sensitivity. If ^{32}P-labeled probes are used, autoradiography is best carried out at –70°C with an intensifying screen whilst ^{35}S-labeled probes are best detected by direct autoradiography. *Figure 6.5* shows a Northern blot analysis of an estrogen-regulated mRNA using a ^{32}P-labeled cDNA probe.

References

1. Watson, J.D., Tooze, J. and Kurtz, D.T. (1983) *Recombinant DNA Technology: A Short Course*. Scientific American Books, New York.

2. Sanger, F. (1981) *Science,* **214**, 1205–1210.

3. Dunn, M.J. (1986) *Gel Electrophoresis of Proteins.* Wright, Bristol.

4. Andrews, A.T. (1986) *Electrophoresis: Theory, Techniques, and Biochemical and Clinical Applications.* Clarendon Press, Oxford.

5. Smith, L.M., Sandos, J.Z., Kaiser, R.J., Hughes, P., Dodd, C., Connell, C.R., Heiner, C., Kent, S.B.H. and Hood, L.E. (1986) *Nature,* **321**, 674–679.

6. Maxam, A.M. and Gilbert, W. (1980) *Methods Enzymol.,* **65**, 499.

7. Williams, G.M., Laspia, M.F. and Dunker, V.C. (1982) *Mutat. Res.,* **97**, 359–370.

8. Butterworth, B., Ashby, J., Bermudez, E., Casciano, D., Mirsalis, J., Probst, G. and Williams, G. (1987) *Mutat. Res.,* **189**, 123–133.

9. Steinmetz, K.L. and Mirsalis, J.C. (1984) *Environ. Mutagen.,* **6**, 321–330.

10. Tyson, C.K. and Mirsalis, J.C. (1985) *Environ. Mutagen.,* **7**, 889–899.

11. Burlinson, B. (1989) *Carcinogenesis,* **10**, 1425–1428.

12. Waters, R. (1984) in *Mutagenicity Testing: A Practical Approach* (S. Venitt and J.M. Parry, eds). IRL Press, Oxford, p. 99–117.

13. Department of Health Publication (1989) *Guidelines for the Testing of Chemicals for Mutagenicity.* HMSO, London.
14. Nose, K. and Okamota, H. (1983) *Biochem. Biophys. Res. Commun.,* **111**, 383–389.
15. Synder, S.D. and Matheson, D.W. (1985) *Environ. Mutagen.,* **7**, 267–279.
16. Manoharan, K., Kinder, D. and Banerjee, M.R. (1987) *Cancer Biochem. Biophys.,* **9**, 127–132.
17. Libbus, B.L., Ilenye, S.A. and Graighead, J.E. (1989) *Cancer Res.,* **49**, 5713–5718.
18. Maehara, Y., Anai, H., Sakaguchi, Y., Kusumoto, T., Emi, Y. and Sugimachi, K. (1990) *Oncology,* **47**, 282–286.
19. Davies, K.E. (1986) *Human Genetic Diseases.* IRL Press, Oxford.
20. Haase, A., Brahic, M., Stowring, L. and Blum, H. (1984) *Methods Virol.,* **3**, 189–226.
21. Cumming, R. and Fallon, R. (1990) in *Radioisotopes in Biology: A Practical Approach* (R.J. Slater, ed.). IRL Press, Oxford, p. 207–242.
22. Southern, E.M. (1975) *J. Mol. Biol.,* **98**, 503–517.

7 Radioisotopes in Nuclear Medicine

In nuclear medicine, an enormous array of different radioisotopes are administered to patients in a variety of pharmaceutical preparations. Such radiolabeled pharmaceuticals are known simply as radiopharmaceuticals. Administration of radiopharmaceuticals invariably has one of two goals; diagnosis or therapy. Both approaches rely on specifically targeting the radiopharmaceutical to the diseased organ or tissue. Diagnostic nuclear medicine relies mostly on imaging the diseased organ (see Section 3.4) although some non-imaging tests are carried out. Therapeutic nuclear medicine relies on using the radiation from the radiopharmaceutical to destroy diseased tissue.

7.1 Radiopharmacology

7.1.1 Choice of radioisotope

Because of the diversity and range of diagnostic imaging studies performed in nuclear medicine, there will never be one radioisotope that is suitable for all applications. Ideal characteristics of radioisotopes for nuclear medicine are multifactorial and include the following [1].

(1) The radioisotope should be a pure γ-emitter with no particulate emissions.

(2) It should decay with a suitable energy of emission for detection by one of the methods described in Section 3.4 (preferred range 100–250 keV).

(3) The radioisotope should be chemically reactive, allowing its incorporation into pharmaceutical molecules.

(4) The eventual radiopharmaceutical should have a suitable effective half-life, that is, a combination of its radioactive and biological half-lives. In addition, it should be non-toxic and preferably carrier free.

(5) The radioisotope should be readily available. Generator produced radioisotopes, which can be produced on site according to demand, are preferable to cyclotron produced radioisotopes.

(6) The radioisotope should be competitively priced, especially where other imaging modalities such as ultrasound and X-ray are capable of producing similar information. Cyclotron produced radioisotopes tend to be more expensive.

Appendix B shows the half-lives, modes of decay and energies of emitted particles/rays of the radioisotopes commonly used for nuclear medicine imaging.

7.1.2 Mechanisms of organ targeting

Diagnostic radiopharmaceuticals exert almost no pharmacological effect with the tracer doses administered. Pathology can be recognized using radiopharmaceuticals by three distinct methods.

(1) Recording areas in a tissue of increased uptake of the radiopharmaceutical within an area of relative homogeneous distribution. Such a situation occurs with tumors due to their increased metabolic activity.

(2) Recording areas in a tissue of reduced uptake within an area of otherwise normal uptake, for example, in necrosis.

(3) Plotting the arrival and disappearance of the radiopharmaceutical over an area of interest, giving functional parameters.

Radiopharmaceuticals have diverse physico-chemical properties and their biodistribution depends upon several factors, for example, their physical state (suspension or solution) or on their metabolic processing in the body. Until recently the majority of uptake mechanisms were poorly understood, especially with ^{99m}Tc-labeled radiopharmaceuticals, where chemical characterization of technetium complexes is difficult. However, radiopharmaceuticals can be targeted to specific sites by one of seven different mechanisms.

(a) Direct localization. This is achieved by the increased uptake of the radiopharmaceutical into the pathological area leading to a 'hot spot' image. An example of this is the simple diffusion of $^{99m}TcO_4^-$ (pertechnetate) along a concentration gradient allowing it to cross a breakdown in a permeability barrier. Thus, $^{99m}TcO_4^-$ will leak across breaches in the blood–brain barrier caused by either a tumor or metastatic lesion or as the result of a cerebrovascular accident.

(b) Indirect localization. This is achieved by using the various physiological and biochemical mechanisms which normally accumulate the radiopharmaceutical in healthy tissue. Where pathology has altered the normal cellular architecture, for example, in necrotic or ischemic myocardium, then uptake does not occur and results in a 'cold spot' image. Various physiological mechanisms are involved in this process and include the following.

(i) *Active transport.* The radiopharmaceutical is taken up against a concentration gradient, for example, [^{99m}Tc]methylene diphosphonate uptake into bone tissue.

(ii) *Phagocytosis.* Particulate substances of diameter 20–500 nm are taken up by the cells of the reticulo-endothelial system due to their recognition by receptors on the cell surface. Opsonization of the particles may occur and phagocytosis then takes place, for example, [^{99m}Tc]sulfur colloid for liver scanning.

(iii) *Capillary blockade.* A particulate suspension of diameter 20–90 μm given by injection will become trapped in the first capillary bed encountered. Thus, when given by the intravenous route, ^{99m}Tc-labeled albumin microspheres or macroaggregates will block blood flow in the pulmonary pre-capillaries (approximate diameter 8 μm), enabling a perfusion lung scan to be performed.

(iv) *Cell sequestration.* The spleen is the major organ that removes damaged erythrocytes from the circulation. Thus, injection of ^{99m}Tc-labeled denatured erythrocytes enables this organ to be visualized.

(c) Compartmental localization. Here, the initial spatial distribution of the radiopharmaceutical is restricted to the fluid compartment. Pathology can then be recognized by imaging, for example, ^{99m}Tc-labeled erythrocytes are used for blood pool imaging, while [^{125}I]human serum albumin is used for plasma volume estimation.

(d) Gases and aerosols. Chemically and physiologically inert gases such as ^{81m}Kr and ^{133}Xe can be used to image the ventilation pathways of the lungs. These gases exchange freely between blood and tissues. Similarly, aerosols generated with a mean particle diameter of 0.5 μm will behave as gases (pseudogases) for short periods of time before the droplets coalesce, and may be used to image lung ventilation as above. After coalescing, the enlarged droplets deposit upon capillary and alveoli walls where, depending on their form (solid or liquid) and physico-chemical properties, they may diffuse into the pulmonary circulation.

(e) Metabolic localization. Positron- (β^+-) emitting radioisotopes of carbon, nitrogen and oxygen can be incorporated into naturally occurring metabolic substances or their analogs, thus enabling fundamental imaging at a true physiological level, for example, glucose metabolism can be studied with [^{11}C]glucose. However, the majority of studies with radiolabeled metabolites have involved the incorporation of 'foreign' radioisotopes into naturally occurring molecules or metabolic analogs, for example, [^{75}Se]selenomethionine or [^{123}I]fatty acids.

(f) Receptor localization. Nuclear medicine is making increasing use of the interaction between radiolabeled compounds and receptors to quantify tissue receptor populations and monitor drug therapy. For example, [^{123}I]iodobenzamide is used to quantify D2-dopamine receptors in the central nervous system and [^{99m}Tc]neogalactosyl albumin is used to quantify asialoglycoprotein receptors in the liver.

(g) Antigenic localization. Monoclonal antibodies recognize a single antigen (see Section 4.1.3). Thus, when radiolabeled, antibodies are taken up by specific cell types, for example, tumors, and can be used for imaging.

7.1.3 Factors affecting biodistribution

Demonstration of the biodistribution of a radiopharmaceutical is only of value in diagnosis if a clear and reliable distinction can be drawn between the normal and diseased state. Due to the inherent nature of radiopharmaceuticals, their biodistribution may be dramatically altered by factors other than the disease process. Such factors include the methods of production and introduction of the radiopharmaceutical into the body and the interaction of the radiopharmaceutical with therapeutically co-administered drugs. For example, therapeutic levels of a drug could swamp or enhance the normal physiological uptake mechanisms for a radiopharmaceutical, leading to false negative or positive results. Clearly, such factors need to be taken into account when interpreting results [2].

7.2 Organ imaging (scintigraphy)

Nuclear medicine studies are unique in their ability to provide functionally-based rather than anatomically-based information. Abnormal anatomy is probably best investigated by conventional X-ray, computed tomography or ultrasound whilst physiological abnormalities are more appropriate to radioisotope studies. The vast majority of these are imaging tests with a small minority being non-imaging.

A thorough critique of the whole range of nuclear medicine studies is obviously outside the range of this chapter, but in this section we will consider the rationale of some of the commonly available imaging studies performed under the relevant organ systems. Section 7.3 will consider some of the non-imaging tests undertaken. Clinical examples of images produced are presented where appropriate. In the clinical environment these are displayed on X-ray film or the γ-camera may be interfaced with a dedicated computer to allow computerized analysis of the resulting data (see Section 3.4).

7.2.1 Central nervous system

The 'classical' brain scan has now been largely superseded by computed tomography. However, pathology can be recognized by direct localization of radiopharmaceuticals such as [^{99m}Tc]DTPA (diethylenetriamine pentaacetic acid) or $^{99m}TcO_4^-$, which cross the damaged blood–brain barrier. Regional cerebral blood flow studies can be performed by positron emission tomography (PET) using endogenous metabolites or their analogs labeled with β^+-emitting isotopes, for example, [^{18}F]fluorodeoxyglucose, [^{15}O]oxygen or [^{13}N]ammonia. The most common PET brain study is with the glucose analog [^{18}F]fluorodeoxyglucose. This radiopharmaceutical is used as a tracer for the exchange of glucose between plasma and brain, and for glucose phosphorylation by hexokinase in this tissue. Application of a mathematical model enables local cerebral glucose metabolism to be calculated [3].

For radiopharmaceuticals to cross the undamaged blood–brain barrier, so enabling γ-camera imaging, they need to be uncharged, lipophilic and of low molecular weight. Once they have crossed the barrier, they must remain fixed for a long enough period of time to allow acquisition of data. ^{123}I-labeled amines such as [^{123}I]iodoamphetamine were originally used [4], but have now been largely overtaken by ^{99m}Tc-labeled ligands such as HMPAO (hexamethylene propyleneamine oxime, see *Figure 7.1*) [5]. This compound exists as two stereoisomers. The *d,l*-isomer is retained in brain tissue whilst the meso-isomer washes out. It has been suggested that this difference in biodistribution may be explained by a difference in interaction with glutathione. Thus, the *d,l*-isomer, having crossed the blood–brain barrier, interacts with free thiol groups in the tissue and is retained, whilst the meso-isomer is free to diffuse out of the tissue again [6]. *Figure 7.2* shows two typical brain images obtained with [^{99m}Tc]HMPAO.

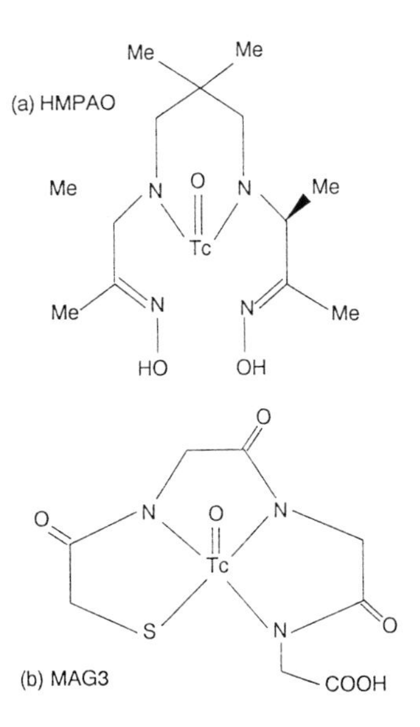

***FIGURE 7.1:** Structural formulae of some recently developed technetium complexes used as radiopharmaceuticals. **(a)** d,l-HMPAO (hexamethylene propyleneamine oxime). **(b)** MAG-3 (mercaptoacetylglycylglycylglycine).*

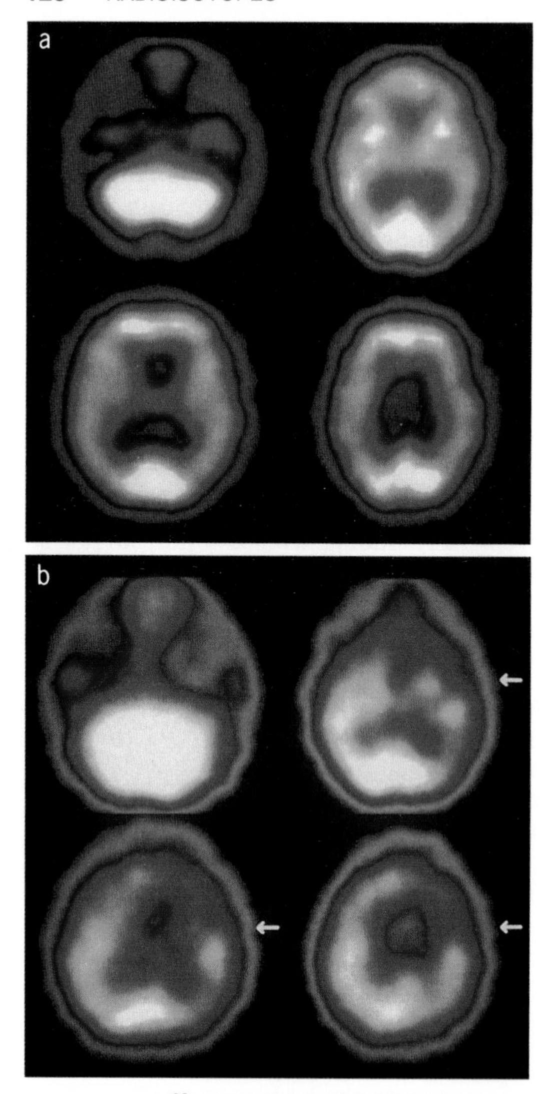

*FIGURE 7.2: [^{99m}Tc]HMPAO SPECT brain images. (**a**) Normal distribution in transaxial slices showing cerebellum (top left), frontal, temporal and occipital lobes (top right), temporal and occipital lobes (bottom left), and parietal lobes (bottom right).(**b**) Same transaxial slices showing reduced perfusion (arrowed) with the tracer in the right temporal lobe (top right and bottom left) and the right parietal lobe (bottom right).*

7.2.2 Cardiovascular system

Three types of clinical investigation are commonly performed.

(a) Myocardial perfusion. Delineation of ischemic viable myocardium is made possible by imaging the left ventricle both at rest and under stress using the radiopharmaceutical [^{201}Tl]thallous chloride [7]. The uptake of $^{201}Tl^{+}$ by myocardial cells is similar to that of K^{+}. Both are monovalent cations and have similar ionic sizes. In order to penetrate into heart muscle cells, the Tl^+ ion must cross the capillary network of the supplying vessels and the interstitial space, and then be taken up across the cell membrane. Thus, both a good coronary blood flow and a functional Na^+/K^+-stimulated ATPase are required to ensure removal of the cation from the blood.

Another radiopharmaceutical used to assess myocardial perfusion is [^{99m}Tc]hexakis-2-methoxyisobutylisonitrile (sestamibi). This gives similar clinical results to ^{201}Tl but has a different uptake mechanism. Again, uptake is primarily related to blood flow, but this lipophilic cationic molecule is retained in heart muscle cells by intracellular binding to a protein of molecular weight 10 kd [8]. *Figure 7.3* shows typical heart images obtained with [^{99m}Tc]sestamibi.

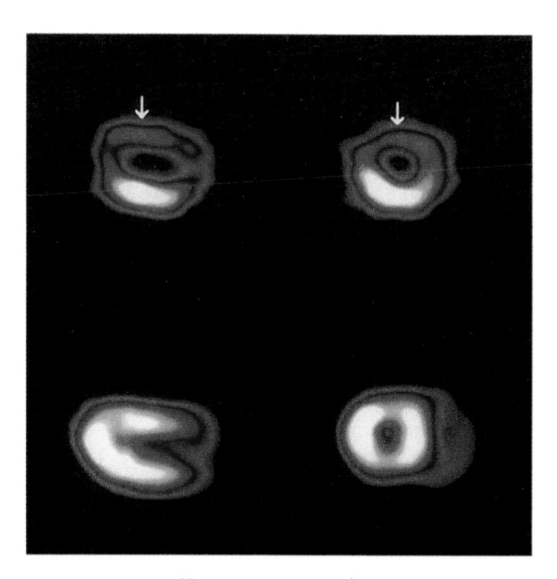

FIGURE 7.3: *[^{99m}Tc]Sestamibi SPECT heart images. The lower two images show normal myocardial perfusion in both the long (left) and short (right) axes. The upper two images show reduced uptake of the tracer (arrowed).*

The PET radiopharmaceutical used to assess myocardial perfusion is ^{82}Rb which can be produced in a generator from the long-lived parent ^{82}Sr. The Rb^+ cation behaves in the same manner as K^+, and is an early marker for impaired heart cell membrane function and/or necrosis [9].

(b) Infarction imaging. Dead or damaged heart muscle cells release their contents as the cells break down. Mitochondria released in this manner become fragmented and calcified, and appear apatite-like on the surface of cells. Since they are extracellular, these calcium deposits are now available for binding with bone-seeking radiopharmaceuticals. Thus, [^{99m}Tc]pyrophosphate is useful for imaging this condition since it complexes with these extracellular calcium deposits [10].

In much the same way, damage to myofibrils in heart muscle causes the extracellular release of myosin. Antimyosin, a monoclonal antibody specific for cardiac myosin, binds to the damaged tissue by antibody/antigen interactions and, when labeled with ^{111}In, can be used to image infarction [11].

(c) Functional studies. Nuclear medicine techniques will give information about cardiac output, for example, the volume of blood ejected by the left ventricle per heart beat. Erythrocytes labeled with ^{99m}Tc (by stannous reduction of the β-hemoglobin chain) are often used [12]. Imaging over the heart will delineate the left ventricle when it is maximally dilated and contracted. Calculations can be performed to give a numerical indication as to how much the heart is functioning (i.e. ejection fraction) [13].

7.2.3 Hepatobiliary system

The superior anatomical resolution of computed tomography, magnetic resonance imaging and ultrasound compared to scintigraphy means that these techniques are able to detect small solitary lesions within the liver. However, since most patients have multiple lesions, scintigraphy and these other imaging modalities give approximately similar information [14].

The basis of liver scanning is that intravenous injection of a ^{99m}Tc-labeled colloid (e.g. tin, sulfur, albumin) will encounter the reticuloendothelial cells which form the framework of the liver and spleen. The diameter of the colloidal particles must be sufficiently small (<0.5 μm) as not to cause capillary blockade. The ^{99m}Tc-labeled colloids are phagocytosed by the Kupffer cells of the liver and remain in the cells for several minutes while imaging can be undertaken. This indirect localization mechanism is thus dependent upon functional reticuloendothelial cells and is particularly useful for detecting diffuse liver disease.

Biliary tract imaging (cholescintigraphy) is performed using various ^{99m}Tc-labeled iminodiacetic acid derivatives (e.g. EHIDA and DISIDA). The HIDA molecule is structurally related to lignocaine and is taken up into hepatocytes by an active transport process. The mechanism of this sinusoidal transport process is poorly understood, but is related to the lipophilicity of the permeant, which must be of molecular weight >300 d, and preferably possess an aromatic ring in its structure. HIDA compounds are secreted across the canalicular membrane into the forming bile by a similar mechanism to that for bilirubin; thus, the two compete for the same carrier [15]. Severely jaundiced patients will have altered biodistribution of HIDA compounds and increased urinary excretion. After biliary excretion, these compounds are eventually excreted in the feces with little or no enterohepatic circulation.

7.2.4 Gastrointestinal tract

Imaging of the gastrointestinal tract is primarily aimed at detecting abnormal function due to disturbed physiology. Anatomical detail is relatively poorly defined compared to other imaging modalities. Some examples of functional and morphological investigations of the gastrointestinal tract are as follows.

(a) Functional tests. Measurement of the rate of transit of a bolus of radioactivity from one part of the gastrointestinal tract to another will give physiological information and parameters which can be compared to a normal range of values.

Esophageal transit time (i.e. the rate of clearance of a bolus of radioactivity from the esophagus into the stomach) and the rate of gastric emptying from the stomach into the small bowel are often measured [16,17]. The boluses of radioactivity may be either liquid or solid in nature, and be a true food or simulated food particles. If part of a liquid meal, the radiopharmaceuticals are usually chelates of DTPA (e.g. ^{99m}Tc, ^{111}In, ^{113m}In) or, if part of a solid phase meal, are usually ^{99m}Tc- labeled colloids [18, 19].

(b) Morphological tests. In order to investigate inflammatory bowel disease, radiolabeled neutrophils are re-injected into the patient and images taken at suitable time intervals over the abdomen to follow cellular migration. The rationale for this test is that factors released by the inflammatory process in affected tissues initiate phagocytosis in the neutrophils, which actively migrate to the areas of inflamed bowel. The radiolabeling of neutrophils is time consuming and requires dedicated equipment and staff. This is due to the lack of a specific radiopharmaceutical for neutrophils. Those presently used are non-selective and indiscriminately label all blood cells. Thus, neutrophils must first be isolated under aseptic conditions from the patient's blood by density gradient centrifugation, and then labeled with a neutral, lipophilic radiopharmaceutical such as [^{99m}Tc]HMPAO or ^{111}In-labeled oxine [20].

Two different radiopharmaceuticals are used to investigate gastrointestinal bleeding and the choice is often related to the urgency of the situation [18]. Normally, ^{99m}Tc-labeled colloid disappears rapidly from the circulation into the reticuloendothelial system (see Section 7.2.3). However, if there is an active gastrointestinal bleed at the time of injection, then any radioactivity circulating in the blood can leak at the site of hemorrhage giving an extra 'hot spot' to the normal reticuloendothelial image. Alternatively, sites of bleeding will produce 'hot spots' in otherwise inactive areas of the bowel after injection of ^{99m}Tc-labeled erythrocytes.

7.2.5 Skeletal system

Bone scans are the single most commonly performed investigation in nuclear medicine world-wide, and account for some 30% of all studies. The radiopharmaceuticals used are the ^{99m}Tc-labeled diphosphonates, the most common being methylene diphosphonate (MDP) [21]. Whilst the mechanism of skeletal uptake is not completely understood, it is partly related to blood flow, as areas of high vascularity show increased uptake. The major influence appears to be osteoblastic activity; diphosphonates accumulate on newly deposited bone surfaces, being incorporated into the forming hydroxapatite crystals rather than the collagenous matrix. Diphosphonates do not affect the nucleation of calcium phosphate but do inhibit the conversion of the amorphous form to hydroxyapatite. Other factors influencing uptake are capilliary permeability, pH balance, fluid pressure within bone and the total mineral content of bone. Thus, uptake is non-specific and occurs in a wide variety of pathologies. The basis of the use of bone scans is that virtually all lesions within the skeleton cause an increase in local osteoblastic activity and an increase in vascularity. Thus, these areas will take up increasing amounts of radioactivity compared to the surrounding area. *Figure 7.4* shows typical bone scans using [^{99m}Tc]MDP.

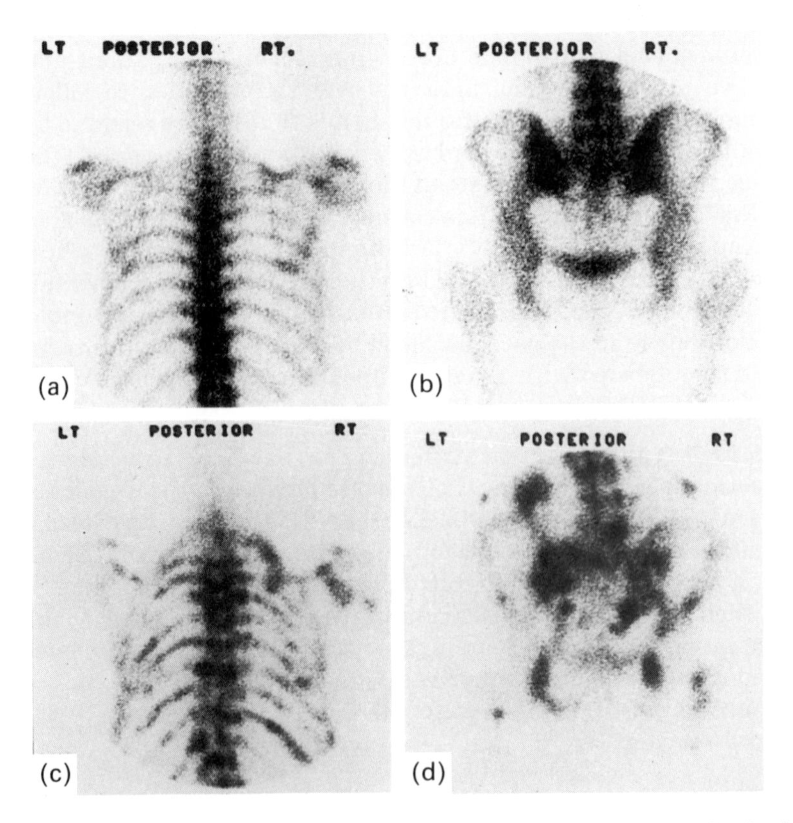

FIGURE 7.4: *[^{99m}Tc]MDP bone images. (**a**) and (**b**) show normal distribution, (**c**) and (**d**) show increased uptake ('hot spots') in areas of metastatic diseases.*

7.2.6 Urinary tract

The technique of renography (i.e. monitoring the passage of a radiopharmaceutical through the kidneys), can give an assessment of individual and/or total function. Morphological information may be obtained by static imaging of the kidneys. Many radiopharmaceuticals have been used and the most common ones are described below.

(a) Dynamic renal imaging agents. When given by intravenous injection, [^{99m}Tc]DTPA is rapidly distributed throughout the extracellular space and is excreted from the body solely by glomerular filtration. It has little protein binding, but this does increase with decreasing renal function [22]. ^{131}I- and ^{123}I-labeled hippuran and [^{99m}Tc]MAG3 (mercaptoacetylglycylglycylglycine, see *Figure 7.1*) are actively secreted by the kidney tubules (about 80%) with the remainder being filtered by the glomeruli [23]. [^{99m}Tc]Glucoheptonate is cleared from the circulation by glomerular filtration alone, but some is reabsorbed by the kidney tubules to be retained in the renal cortex.

(b) Static renal imaging agents. After intravenous injection, [^{99m}Tc]DMSA (dimercaptosuccinic acid) is extensively protein bound and is filtered slowly by the glomerulus. In addition, it is reabsorbed from the renal tubules. Thus, like glucoheptonate, it accumulates in the renal cortex and gives a high specificity of uptake [24].

7.2.7 Endocrine system

The function of several endocrine glands can be studied using radiopharmaceuticals and some of these are described below.

(a) Thyroid gland. The most commonly used radiopharmaceutical to image the thyroid gland is sodium [^{99m}Tc]pertechnetate (see *Figure 7.5*).

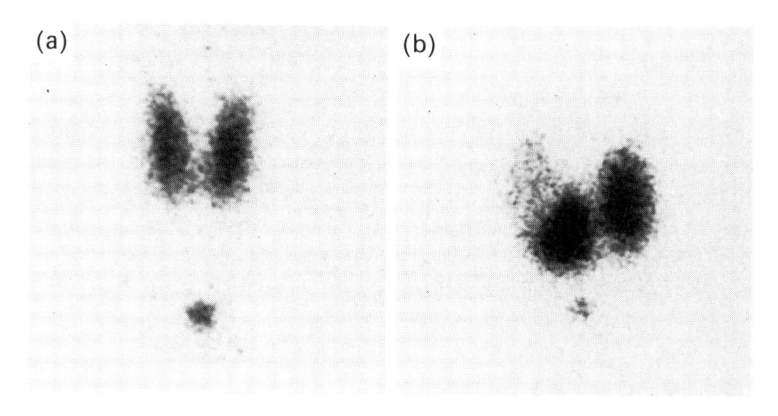

FIGURE 7.5: *Sodium [^{99m}Tc]pertechnetate thyroid images. (**a**) Normal distribution. The 'hot spot' at the bottom of the image is a marker to give positional information. (**b**) Increased uptake in a patient with nodular goiter.*

The TcO_4^- ion is taken up into thyroid follicle cells by the same active transport system as for I^-; this is because they are of similar ionic weight and charge. However, TcO_4^- is not incorporated into tyrosine residues of thyroglobulin (as happens with I^- during the biosynthesis of thyroxine). $^{131}I^-$ and $^{123}I^-$ are also used for imaging purposes [25]; these are taken up by active transport and incorporated into thyroxine. Thus, both radioisotopes can be used to quantify uptake by thyroid follicular cells.

(b) Parathyroid glands. These glands cannot be imaged directly due to the much larger and overlying thyroid gland. However, the overactive glands can be imaged by subtraction imaging using two different radioisotopes with different energies of γ-emission. Both parathyroid and thyroid tissue take up [^{201}Tl]thallous chloride by a non-specific mechanism related to cellularity and vascularity. Subsequent injection of $^{99m}TcO_4^-$ is then carried out to give a thyroid-only image. Computer analysis is then used to subtract the thyroid-only image from the combined parathyroid and thyroid image to leave a pattern of distribution of the hyperfunctioning parathyroid tissue [26].

(c) Pancreas. The endocrine pancreas can also be imaged by a subtraction technique. A radiolabeled amino acid analog, [^{75}Se]selenomethionine, is imaged after uptake into liver and pancreas [27]. Uptake occurs due to the high turnover of proteins in these tissues. The unwanted liver uptake on this image is subtracted by performing a ^{99m}Tc-labeled colloid scan as described in Section 7.2.3.

(d) Adrenal glands. Corticosteroids are synthesized in the adrenal cortex from cholesterol derived from cholesterol esters stored in the cortex. When administered intravenously, radiolabeled cholesterol analogs are concentrated in this cholesterol ester pool. Thus, examples of radiopharmaceuticals used to image the adrenal cortex are 6-[^{131}I]iodomethyl-19-norcholesterol and 6-[^{75}Se]selenomethyl-19-norcholesterol [28]. The rate of uptake and degree of concentration in the cortex is variable, and depends largely on the pathological condition of the patient. Subtraction imaging of the kidneys may be necessary to assist localization.

The cells of the adrenal medulla can be considered as specialized post-ganglionic sympathetic neurones and can also be imaged with radiolabeled adrenergic blocking agents which localize in this tissue and other sympathetic nerve endings. The radiopharmaceutical used for diagnostic imaging is [^{123}I]metaiodobenzylguanidine; this agent is effectively part of the adrenergic blocking drug bretylium and a substituted guanidine moiety. Uptake into the adrenal medulla is via a neuronal pump [29].

7.2.8 Infection and inflammation imaging

The use of radiopharmaceuticals to image these processes assumes that localization to inflammation sites is due to the cellular response to

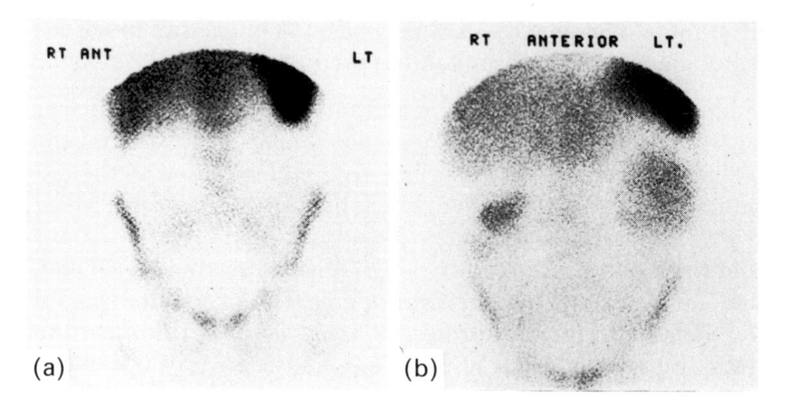

FIGURE 7.6: *[^{111}In]-labeled neutrophil imaging. (**a**) Normal distribution. (**b**) Left sided uptake in an abscess.*

infection and is therefore non-specific in nature. Thus, as described in Section 7.2.4, radiolabeled neutrophils will migrate and accumulate in areas of infection [30]. *Figure 7.6* shows the localization of an abscess by ^{111}In-labeled neutrophils.

[^{67}Ga]Gallium citrate is also used for imaging infection. This accumulates at sites of inflammation for two principal reasons. Because Ga^{3+} is similar to Fe^{3+}, it binds to transferrin (an iron transport protein in plasma) which leaks non-specifically from capillaries surrounding sites of inflammation [31]. Ga^{3+} also binds to lactoferrin liberated from neutrophils during bacterial destruction.

7.2.9 Tumor imaging

Many radiopharmaceuticals are taken up by a variety of tumors, mostly by non-specific methods. Accumulation in transformed cells is primarily due to changes in blood supply and metabolism. The mechanisms of accumulation of radiopharmaceuticals by tumors can be broadly classified as either substrate non-specific or substrate specific.

(a) Non-specific tumor imaging radiopharmaceuticals. Many of the ^{99m}Tc-labeled radiopharmaceuticals can be used in this respect. Tumors and metastatic tissue in the early stages of their development have an increased blood supply giving increased leakage of $^{99m}TcO_4^-$ from capillaries. Similarly, $^{67}Ga^{3+}$ bound to transferrin also shows increased leakage from capillaries into tumors. In addition, because growth rate is faster in tumors than in normal tissue, rates of metabolism are invariably faster. Thus, for example, bone tumors can be imaged as areas of increased uptake of [^{99m}Tc]methylene diphosphonate.

(b) Specific tumor imaging radiopharmaceuticals. A classic example of this is increased uptake of $^{131}I^-$ or $^{123}I^-$ by thyroid tumors.

Similarly, [^{123}I]metaiodobenzylguanidine is used to image tumors of the adrenal medulla and other catecholamine- producing tumors.

Radiolabeled monoclonal antibodies are finding increasing use in nuclear medicine for tumor imaging. The rationale for this approach is that the antibody is directed towards an abnormal antigen expressed at the cell surface of tumor tissue [32]. Of the five classes of antibody, the commonest types used for antibody imaging are IgGs and IgMs. Many of the monoclonal antibodies used for tumor imaging are of murine origin and, when injected into man, are themselves antigenic. Thus, repeated studies in a single patient typically produces increased uptake by the reticuloendothelial system (liver and spleen), and decreased uptake by the tumor.

At the present time, several monoclonal antibodies are undergoing multicenter clinical trials for tumor imaging, and, in the future, it is possible that immunoscintigraphy will become the method of choice for tumor imaging [33].

7.3 Non-imaging tests

Nuclear medicine studies involving non-imaging techniques are diverse in nature. However, they have the following features in common:

(1) Because they involve machine counting of biological samples (blood, urine, feces, exhaled air, etc.) rather than imaging, the dose of radioactivity given to a patient can be kept much lower than that needed for imaging.

(2) They utilize radiopharmaceuticals labeled with long-lived radioisotopes (e.g. ^{14}C, ^{51}Cr, ^{57}Co, ^{125}I) which are radiochemically stable.

(3) Non-imaging tests allow precise quantitation of results as d.p.m. or c.p.m. per unit of sample. When samples are taken at known time intervals, time–activity curves can be generated and compared to control curves. If samples are not collected, residual body activity can be monitored by surface counting methods or by means of a whole body counter.

Two commonly used non-imaging tests are the determination of circulating erythrocyte volume using ^{51}Cr-labeled erythrocytes, and the determination of plasma volume using ^{125}I-labeled human serum albumin. Both these tests use isotope dilution analysis, the principle of which is as follows. A suspension/solution of the radiolabeled tracer (containing C_0 c.p.m./ml in V_0 ml) is injected intravenously. This will become distributed into the unknown volume, V_x, such that the concentration of radioactivity is diluted to C_i c.p.m./ml in a final volume of $(V_x + V_0)$ ml. Thus, by measuring the radioactivity in a sample taken after administration and thorough mixing of the tracer, the unknown volume can be calculated from the equation,

$$V_x = V_0(C_0/C_i - 1).$$

If the injected volume (V_0) is very small compared to V_x (as is usually the case), the dilution on injection can be ignored and the above equation is simplified to,

$$V_x = V_0(C_0/C_i).$$

(In addition to determining volumes of distribution, isotope dilution analysis can be used to measure accurately small quantities of compounds providing the compound is available in a radiolabeled form. A full discussion of radioisotope dilution analysis can be found in reference [34].)

Table 7.1 describes some of the non-imaging tests commonly performed in nuclear medicine.

TABLE 7.1: *Some commonly performed non-imaging tests in nuclear medicine*

Radiopharmaceutical	Clinical indication	Administration route	Sample collected
[^{51}Cr]Erythrocytes [35]	Erythrocyte mass	IV	Blood
	Erythrocyte survival	IV	Blood
	Gastrointestinal blood loss	IV	Feces
[^{51}Cr]Chromic chloride [36]	Gastrointestinal protein loss	IV	Feces
[^{51}Cr]EDTA [37]	Glomerular filtration rate	IV	Plasma
[^{125}I]Hippuran [38]	Effective renal plasma flow	IV	Plasma/ urine
[^{125}I]Human serum albumin [39]	Plasma volume	IV	Plasma
[^{59}Fe]Ferric citrate [40]	Ferrokinetics	IV	Blood
[^{57}Co]Cyano-cobalamin [41]	Pernicious anemia	Oral (capsule)	Urine
[^{75}Se]Taurosel-cholic acid [42]	Bile acid pool loss	Oral (capsule)	Feces
[^{14}C]Cholic acid [43]	Bile acid absorption	Oral (drink)	Breath (CO_2 trap)
[^{14}C]Triolein [44]	Fat absorption	Oral (drink)	Breath (CO_2 trap)
[^{14}C]Aminopyrine [45]	Liver function	Oral (drink)	Breath (CO_2 trap)

7.4 Radioisotope therapy

Radioisotopes have been used therapeutically for the treatment of disease for at least 50 years. Original applications were in the fields of endocrinology and hematology. More recently, interest has centered on rheumatology and currently oncology has come to the fore with many specific radiopharmaceuticals being used for both the diagnosis and treatment of tumors. Indeed, it has been shown that there is a much lower risk of leukemia and secondary tumors from this form of treatment than from chemotherapy or radiation therapy from an external source [46].

For successful therapy, the radiopharmaceutical must have a long retention time and be highly selective to avoid damage to normal tissue. Its effect will depend upon the sensitivity of the lesion to radiation and upon the absorbed radiation dose-equivalence.

The most common approach is to use a low energy β-emitter such that the range of the β-particles is compatible with the distance of the target (cell nucleus or membrane) from the point of attachment (if known) of the radiopharmaceutical at the site of uptake. Other approaches are to use high energy β- or α-emitters which give higher doses over a short path length, or Auger electron-emitting radioisotopes if attachment to the cell nucleus is possible. *Table 7.2* shows the physical characteristics of some radioisotopes used for therapy.

TABLE 7.2: *Physical characteristics of some radioisotopes used in therapy*

Radioisotope	Half-life	Emission	Max. energy (MeV)	Max. range
^{32}P	14.3 d	β	1.71	8.7 mm
^{67}Cu	2.6 d	β and γ	0.58	2.2 mm
^{89}Sr	50.5 d	β	1.5	8.0 mm
^{90}Y	2.7 d	β	2.3	12.0 mm
^{125}I	60.0 d	Auger	—	10.0 nm
^{131}I	8.0 d	β and γ	0.61	2.4 mm
^{153}Sm		2.0 d	β and γ	0.81 3.0 mm
^{165}Dy	2.3 h	β and γ	1.3	6.4 mm
^{169}Er	9.5 d	β	0.34	1.0 mm
^{186}Re	3.8 d	β	1.08	5.0 mm
^{198}Au	2.7 d	β and γ	0.96	4.4 mm
^{211}At	7.2 h	α	6.8	65.0 μm
^{212}Bi	1.0 h	α	7.8	70.0 μm

Data reproduced from reference [50] with permission of Professor Hoefnagel and Springer-Verlag, Heidelberg.

The same considerations apply for the uptake of therapeutic radiopharmaceuticals as apply for diagnostic agents (see Section 7.1.2). However, there are further criteria to consider to ensure that the radiopharmaceutical is brought into close proximity with the target cells. One such criterion is that the target cells must be directly accessible to the radiopharmaceutical, that is, if administered intravenously, there should be a good blood supply. If this is not so, the only alternative is direct application of the radiopharmaceutical to the target. Target cells must have sufficient 'binding' sites available to retain the radiopharmaceutical; thus, the radiopharmaceutical should be carrier free. Finally, the effective half-life of the radiopharmaceutical should be due solely to the rate of decay of the radioisotope. If necessary, drugs can be co-administered to alter the biological half-life of the radiopharmaceutical.

7.4.1 Hyperthyroidism and thyroid cancer

Sodium [^{131}I]iodide has been used in thyroid therapy for over 40 years [47]. The rationale for its use is as described in Section 7.2.7. Tracer doses are given to estimate uptake and, from this, therapeutic doses can be calculated. However, it should be noted that large therapeutic doses often have different kinetics from tracer doses.

7.4.2 Metastatic bone disease

Bone metastases can be treated palliatively and with long duration using [^{89}Sr]strontium chloride. Bone is solid tissue formed by the mineralization of its organic structure by calcium phosphate deposited as crystals of hydroxyapatite. This is a dynamic process with constant formation and reabsorption occurring under the control of the hormones calcitonin and parathormone, and is influenced by the pH of interstitial fluids and mechanical pressure.

Like Ca^{2+}, $^{89}Sr^{2+}$ is a Group II element, and when injected intravenously will be deposited in bone as strontium phosphate. Its biodistribution is similar to Ca^{2+} in all respects except that its urinary excretion is 3–8 times higher [49]. Metastatic lesions have increased uptake of $^{89}Sr^{2+}$ due to increased metabolic activity of the transformed osteoblastic cells.

Recently, two further β-emitters, ^{186}Re (rhenium) and ^{153}Sm (samarium) have been chelated to compounds that behave in a similar fashion to the diagnostic bone agent, [^{99m}Tc]methylene diphosphonate. There is good correlation between these diagnostic and therapeutic agents in their bone to lesion ratios, and they produce similar results to $^{89}Sr^{2+}$ [50].

7.4.3 Radioimmunotherapy

Treatment of tumors with monoclonal antibodies labeled with therapeutic radioisotopes offers several potential advantages, not the least of which is specificity of uptake by the target tumor. This can be further improved by local or regional, rather than systemic, administration. However, at present this form of treatment remains in an experimental stage and is undergoing world-wide clinical trials.

References

1. NCRP Report No. 70 (1982) *nuclear medicine – Factors Influencing the Choice and Use of Radionuclides in Diagnosis and Therapy.* NRCP Publications, Bethesda.
2. Sampson, C.B. (1990) *Textbook of Radiopharmacy; Therapy and Practice*. Gordon and Breach, London.
3. Rervich, M. (1979) *Circ. Res.,* **44**, 127–137.
4. Holman, B.L., Lee, R.G.L., Hill, T.C., Lovett, R.D. and Lister-James, J. (1984) *J. Nucl. Med.,* **25**, 25–30.
5. Sharp, P.F., Smith, F.W., Gemmell, H.G., Lyall, D., Evans, N.T.S., Gvozdanovic, D., Davidson, J., Tyrell, D.A., Pickett, R.D. and Nievinck, R.D. (1986) *J.Nucl.Med.,* **27**, 171–177.
6. Ballinger, J.R., Reid, R.H. and Gulinchyn, K.Y. (1988) *J.Nucl.Med.,* **29**, 1998–2000.
7. Wackers, F.J. (1980) *Semin. Nucl. Med.,* **X**, 127–145.
8. Okoale, R.D., Glover, D., Gaffney, T. and Williams, S. (1988) *Circulation,* **77**, 491–498.
9. Mullani, N.A., Goldstein, R.A., Glould, K.L., Fisher, D.S., Marani, S.K. and O'Brein, H.A. (1983) *J. Nucl. Med..* **24**, 898–906.
10. Willerson, J.T., Parkey, R.W., Bonte, F.J., Lewis, S.E., Corbeth, J. and Buja, L.M. (1980) *Semin. Nucl. Med.,* **X**, 54–69.
11. Khaw, B.A., Fallon, J.T., Strauss, H.W. and Haber, E. (1980) *Science,* **209**, 295–297.
12. Smith, T.D. and Richards, P. (1976) *J. Nucl. Med.,* **17**, 126–132.
13. Thrall, J.H., Freitas, J.E., Swanson, D. Rogers, W.L., Clare, J.M., Brown, M.L. and Pitt, B. (1978) *J. Nucl. Med.,* **19**, 796–803.
14. Biello, D.R., Levitt, R.G., Siegel, B.A., Sagel, S.S. and Stanley, R.J. (1978) *Radiology,* **127**, 159–163.
15. Harding, K.L., Sorgi, M., Wolverson, R.L., Mosimann, F., Sherwin, S., Donovan, I.A. and Alexander-Williams, J. (1984) *Scand. J. Gastroenterol.,* **19** (suppl 92), 27–29.
16. Siebert, J.J., Byrne, W.J., Enlet, A.R., Latture, A., Leach, M. and Campbell, M. (1983) *Am. J. Roentgenol.,* **104**, 1087–1090.
17. Donovan, I.A. and Harding, L.K. (1986) in *Nuclear Gastroenterology* (P.J. Robinson, ed.). Churchill Livingstone, Edinburgh, p. 24–35.
18. Harding, L.K. and Robinson, P.J. (1990) *Clinicians Guide to nuclear medicine: Gastroenterology.* Churchill Livingstone, Edinburgh.
19. Harding, L.K. (1989) in *Practical Guide to nuclear medicine* (P.F. Sharp, H.G. Gammel and F.W. Smith, eds.). IRL Press, Oxford, p. 204–205.

20. Danpure, H.T. and Osman, S. (1989) in *Radiopharmaceuticals; Using Radioactive Compounds in Pharmaceuticals and Medicine* (A.E. Theobold, ed.). Ellis Horwood, Chichester, p. 65–82.

21. Subramanian, G., McAfree, J.G., Blair, R.J., Kallfetz, F.A. and Thomas, F.D. (1975) *J. Nucl. Med.,* **16**, 744–755.

22. Piepsz, A., Dobbelan, A. and Erbsman, F. (1977) *Eur. J. Nucl. Med.,* **2**, 173–177.

23. Jafri, R.A., Britton, K.E. and Nummon, C.C. (1988) *J. Nucl. Med.,* **29**, 147–158.

24. Vanlic-Razumenic, N., Malesevic, M. and Stefanovic, L.J. (1979) *Nukl. Med.,* **18**, 40–45.

25. Beierwaltes, W.H. (1990) *J. Nucl. Med.,* **31**, 400–402.

26. Young, A.E., Gaunt, J.L. and Croft, D.N. (1983) *Br. Med. J.,* **286**, 1384–1386.

27. Zuroski, S., Graban, W.T. and Jakubowski, W. (1977) *Eur. J. Nucl. Med.,* **2**, 273–275.

28. Sarkar, S.D., Cohen, E.L., Beierwaltes, W.H., Ice, R.D., Cooper, R. and Gold, E.N. (1977) *J. Clin. Endocrinol. Metab.,* **45**, 353–362.

29. Wieland, D., Wu, J., Brown, L.E., Manger, T.J., Swanson, D.P. and Beierwaltes, W.H. (1980) *J. Nucl. Med.,* **21**, 349–353.

30. Segal, A.W., Arnot, R.N., Thakur, M.L. and Lavender, J.P. (1976) *Lancet,* **2**, 1056–1058.

31. Hoffer, P. (1980) *J. Nucl. Med.,* **21**, 484–488.

32. Kohler, G. and Milstein, C. (1975) *Nature,* **256**, 494–502.

33. Perkins, A.C. and Pimm, M.V. (1991) *Immunoscintigraphy; Practical Aspects and Clinical Applications.* Wiley-Liss, New York.

34. Tolgyessy, J., Braun, T. and Krys, M. (1972) *Isotope Dilution Analysis.* Pergamon Press, Oxford.

35. Gray, S.J. and Sterling, K. (1950) *J. Clin. Invest.,* **29**, 1604–1613.

36. Reba, R.C. and Salkeld, J. (1982) *Semin. Nucl. Med.,* **XII**, 147–155.

37. Brochner-Mortensen, J. (1978) *Dan. Med. Bull.,* **25**, 181–202.

38. Chervu, L.R. and Blaufox, M.D. (1982) *Semin. Nucl. Med.,* **XII**, 224–245.

39. Wright, R.R., Tono, M. and Pollycore, M. (1975) *Semin. Nucl. Med.,* **V**, 63–78.

40. McIntyre, P.A. (1972) *Hosp. Prac.,* **7**, 99–108.

41. Knudsen, L. and Hippe, E. (1974) *Scand. J. Haemat.,* **13**, 287–293.

42. Deltiez, H., Van den Berg, J.W.O., Van Blankenstein, M. and Meerisaldt, J.H. (1982) *Eur. J. Nucl. Med.,* **7**, 269–271.

43. Caspary, W.F. (1978) *Clin. Gastroenterol.,* **7**, 351–374.

44. Turner, J.M., Lawrence, S., Fellows, I.W., Johnson, I., Hill, P.G. and Holmes, G.K. (1987) *Gut,* **28**, 694–700.

45. Hepner, G.W. and Vesell, E.S. (1974) *New Engl. J. Med.,* **291**, 1384–1388.

46. Beierwaltes, W.H. (1985) *J. Nucl. Med.,* **26**, 421–427.

47. Maxon, H.R., Thomas, S.R. and Wen Chen, I. (1981) *Clin. Nucl. Med.,* **6**, 87–88.

48. Blake, G.M., Ziranovic, M.A., McEwan, A.J. and Ackery, D.M. (1986) *Eur. J. Nucl. Med.,* **12**, 447–454.

49. Spencer, H., Laslo, D. and Brothers, M. (1957) *J. Clin. Invest.,* **36**, 680–685.

50. Hoefnagel, C. (1991) *Eur. J. Nucl. Med.,* **18**, 408–431.

Appendix A. Physical Properties of Some Radioisotopes Commonly Used in the Biological Sciences

Element	Radio-isotope	Half-life ($t_{½}$)	Mode of decay	Energy of radiation (MeV)	
				$E_{\beta max}$	E_{γ}
Hydrogen	^{3}H	12.26 y	β^{-}	0.018	—
Carbon	^{14}C	5736 y	β^{-}	0.156	—
Sodium	^{22}Na	2.6 y	β^{+} (90.5%)	0.54	1.28
			EC (9.5%)	—	0.51
	^{24}Na	15.0 h	β^{-}	1.39	1.37, 2.75
Phosphorus	^{32}P	14.3 d	β^{-}	1.71	—
Sulfur	^{35}S	87.4 d	β^{-}	0.167	—
Chlorine	^{36}Cl	3×10^{5} y	β^{-}	0.714	—
Potassium	^{42}K	12.4 h	β^{-}	2.0 (18%)	1.52 (18%)
				3.6 (82%)	
Calcium	^{45}Ca	165 d	β^{-}	0.254	—
Chromium	^{51}Cr	27.8 d	EC	—	0.323 (8%)
Iron	^{55}Fe	2.7 y	EC	—	0.0059 (23%)
	^{59}Fe	45 d	β^{-}	0.27 (46%)	1.10 (57%)
				0.46 (54%)	0.19 (2.4%)
Cobalt	^{57}Co	270 d	EC	—	0.122 (89%)
	^{60}Co	5.26 y	β^{-}	0.31	1.17, 1.33
Iodine	^{125}I	60 d	EC	—	0.035 (7%)
	^{131}I	8.04 d	β^{-}, γ	0.61 (86%)	0.364 (80%)
				0.34 (13%)	0.284, 0.637

Appendix B. Physical Properties of Some Radioisotopes Commonly Used for Nuclear Medicine Imaging

Element	Radio-isotope	Half-life ($t_{1/2}$)	Mode of decay	Energy of radiation (MeV)	
				$E_{\beta max}$	E_{γ}
Carbon	^{11}C	20 min	β^+	0.97	0.51
Nitrogen	^{13}N	10 min	β^+	1.2	0.51
Oxygen	^{15}O	2.0 min	β^+	1.7	0.51
Fluorine	^{18}F	110 min	β^+(97%) EC (3%)	0.649	0.51
Gallium	^{67}Ga	78.2 h	EC		0.09 (42%) 0.18 (24%) 0.30 (22%)
Selenium	^{75}Se	119.8 d	EC		0.14 (54%) 0.26 (56%)
Krypton	^{81m}Kr	13 sec	IT		0.19
Rubidium	^{82}Rb	1.27 min	β^+ γ	3.4	0.51 0.78
Technetium	^{99m}Tc	6.0 h	IT		0.140
Indium	^{111}In	2.8 d	EC		0.173 0.247
	^{113m}In	104 min	IT		0.390
Iodine	^{123}I	13 h	EC		0.159
	^{131}I	8.04 d	β^-, γ	0.61 (86%) 0.34 (13%)	0.364 (80%) 0.284, 0.637
Xenon	^{127}Xe	36.4 d	EC		0.17, 0.20 0.38
	^{133}Xe	5.3 d	β^-, γ	0.34	0.081 (35%)
Thallium	^{201}Tl	72.9 h	EC, γ		0.008 0.170

Appendix C. Further Reading

In this short book we have attempted to describe the basic principles and methods for working with radioisotopes, and their more modern applications. We have tried to present sufficient detail and references for the reader to design new applications of radioisotopes. For those wishing to pursue the subject further, the following may be useful.

Chapman, J.M. and Ayrey, G. (1981) *The Use of Radioactive Isotopes in the Life Sciences*. George Allen and Unwin, London.

Chard, T. (1987) *An Introduction to Radioimmunoassay and Related Techniques*. Elsevier, Amsterdam.

Evans, E.A. and Oldham, K.G. (1988) *Radiochemicals in Biomedical Research: Critical Reports on Applied Chemistry*, Vol. 24. Wiley, Chichester.

Geary, W. (1986) *Radiochemical Methods*. Wiley, Chichester.

Kricka, L.J. (1985) *Ligand-Binder Assays: Labels and Analytical Strategies: Clinical and Biochemical Analysis*, Vol. 17. Marcel Decker, New York.

Parker, R., Smith, P. and Taylor, D. (1978) *Basic Science of Nuclear Medicine*. Churchill Livingstone, Edinburgh.

Sampson, C.B. (1990) *Textbook of Radiopharmacy*. Gordon and Breach, London.

Slater, R.J. (1990) *Radioisotopes in Biology: A Practical Approach*. IRL Press, Oxford.

Sharp, P., Gemmell, H. and Smith, F. (1989) *Practical Nuclear Medicine*. IRL Press, Oxford.

Webb, S. (ed.) (1988) *The Physics of Medical Imaging*. Adam Hilger, Bristol.

Appendix D. Some Useful Addresses

Where possible, addresses in the UK and the USA are given.

(A) Radiochemicals, radiopharmaceuticals and radioimmunoassay kits

Amerlite Diagnostics Ltd	Manderville House, 62 The Broadway, Amersham, Bucks HP7 0HJ, UK	
Amersham International plc	Lincoln Place, Green End, Aylesbury, Bucks HP20 2TP, UK	Amersham Corporation, 2636.5 Clearbrook Drive, Arlington Heights, IL 60005, USA
Becton Dickinson Ltd	Between Towns Road, Cowley, Oxford OX4 3LY, UK	
Behring Ltd	50 Salisbury Road, Hounslow, Middx TW4 6JH, UK	
Bristol Myers Squibb Diagnostics Inc.		PO Box 4500, Princeton, NJ 08543-4500, USA
Celltech Diagnostics Ltd	240 Bath Road, Slough, SL1 4ET, UK	
CIS Ltd	Dowding House, Wellington Road, High Wycombe, Bucks HP12 3PR, UK	10 De Angelo Drive, Bedford, MA 07130-2267, USA

Diagnostic Products Ltd	22 Blacklands Way, Abingdon Business Park, Abingdon, Oxon OX14 1DY, UK	5700 West 96th, Los Angeles, CA 90045, USA
Du Pont Ltd	(Radiochemicals), Wedgewood Way, Stevenage, Herts SE1 4QN, UK	(Radiochemicals), 1007 Market Street, Wilmington, DE 19898, USA
	(Radiopharmaceuticals), Acton Grange District Centre, Birchwood Lane, Warrington, Cheshire WA4 6XE, UK	(Radiopharmaceuticals), Diagnostic Imaging Div., 331 Treble Cove Road, N. Billerica, MA 01862, USA
Malinckrodt Medical	11 North Portray Close, Round Spiney, Northampton NW 3 3RQ, UK	2703 Wagner Place, Maryland Heights, MO 63043, USA
Medgenix Diagnostics Ltd	19 Castle Street, High Wycombe, Bucks HP13 6RU, UK	
Metachem Diagnostics Ltd	29 Forest Road, Piddington, Northampton NN7 3LY, UK	
Serono Diagnostics Ltd	Technitron House, Unit 2, Redfields Business Park, Church Cookham, Fleet, Hants GU13 0RD, UK	Serono Baker Diagnostics Inc., 100 Cascade Drive, Allentown, PA 18103, USA
Sigma Chemical Co. Ltd	Fancy Road, Poole, Dorset BH17 7NH, UK	P.O. Box 14508, St Louis, MO 63178, USA

(B) Scintillation and gamma counters

Beckmann Instruments Ltd	Progress Road, Sands Industrial Estate, High Wycombe, Bucks HP12 4JL, UK	Nuclear Systems Operations, 2500 Harbor Boulevard, Fullerton, CA 92634, USA

Canberra-Packard Ltd	Brook House, 14 Station Road, Pangbourne, Berks RG8 7DT, UK	Packard Instrument Co., 2200 Warrenville Road, Downers Grove, IL 60515, USA
Pharmacia Ltd	Pharmacia House, Midsummer Boulevard, Milton Keynes, MK19 3HP, UK	9319 Gaither Road, Gaithersburg, MD 2877, USA

(C) Autoradiography

Alpha Laboratories Ltd	40 Parham Drive, Eastleigh, Hants SO5 4NU, UK	
Amersham International plc	Lincoln Place, Green End, Aylesbury, Bucks HP20 2TP, UK	Amersham Corporation, 2636.5 Clearbrook Drive, Arlington Heights, IL 60005, USA
Du Pont Ltd	Wedgewood Way, Stevenage, Herts SE1 4QN, UK	1007 Market Street, Wilmington, DE 19898, USA
Genetic Research Instruments Ltd	Gene House, Dunmow Road, Felsted, Essex CM6 3LD, UK	
Kodak Ltd	Health Sciences Div., P.O. Box 66, Station Road, Hemel Hempstead HP1 1JU, UK	Eastman Kodak, 343 State Street, Rochester, NY 14650, USA

(D) Non-film autoradiography

Betagen Corporation	Genetic Research Instruments Ltd, Gene House, Dunmow Road, Felsted, Essex CM6 3LD, UK	100 Beaver Street, Waltham, MA 02154, USA

Appendix E. Legislation Concerning Radioisotopes

Whilst enforceable legislation controlling the use of radioisotopes exists in almost all countries, the emphasis on certain aspects of the law varies considerably from one country to another. In the UK, Parliament passes wide ranging acts, thereby enabling specific regulations to be laid down concerning pertinent aspects of the Act. To enable interpretation of the regulations, Approved Codes of Practice and Guidance Notes may be issued which lay down preferred methods of working.

(1) The Health and Safety at Work Act (1974)

This ensures the general health and safety of people at work and members of the public.

The **Ionizing Radiation Regulations (1985)** are enforceable under this Act and relate to all matters of health and safety with respect to ionizing materials. They cover responsibility for radiation protection, classification of employees, monitoring of staff and the workplace, controlled areas, storage of radioisotopes, accounting of radioactivity and washing and changing facilities. Thus, these regulations contain the fundamental requirements to control exposure to ionizing radiation.

The Ionizing Radiation Regulations **Approved Code of Practice** covers the protection of persons against ionizing radiation arising from any work activity. This document gives details of acceptable methods of meeting the requirements of the Regulations.

Guidance Notes for the protection of persons against ionizing radiations arising from medical and dental use have been prepared by the National Radiological Protection Board. These notes are a guide to good radiation protection practice consistent with the regulatory requirements. They are for use by anyone in medical and dental practice, and in allied research, who uses ionizing radiation in human subjects. They cover irradiation, whether for diagnostic, therapeutic or research purposes, and where *in vitro* medical tests are conducted.

The **Ionizing Radiation (Protection of Persons Undergoing Medical Examination or Treatment) Regulations (1988)** ensure that persons effecting medical exposure are adequately trained.

(2) The Radioactive Material (Road Transport) Act (1991)

This Act enables legislation to be made enforcing conformation to International Transport Regulations.

The **Radioactive Substances (Carriage by Road) Regulations (1974)** cover the transport of radioisotopes by road and includes a **Code Of Practice**.

(3) The Medicines Act (1968)

Radiopharmaceuticals are considered to be medicinal products and are therefore controlled under this Act.

(4) The Radioactive Substances Act (1960)

This Act covers registration of premises, the keeping and use of radioisotopes, and their subsequent accumulation and/or waste disposal. Her Majesty's Inspectorate of Pollution issue certificates of authorization detailing limits of radioactivity to be kept, routes and amounts of disposal and record keeping requirements.

All workers with radioisotopes in the UK are governed by these various acts and regulations. Workers outside the UK are strongly advised to familiarize themselves with their own national legislation before commencing work with radioactive materials.

Appendix F. Glossary

Alpha (α) particle: the nucleus of a helium atom (two protons plus two neutrons) emitted by radioactive isotopes whose atomic number is >82.

Annihilation reaction: the mutual destruction of an electron–positron pair resulting in the production of two γ-rays, each with an energy of 0.51 MeV.

Antibody: an immunoglobulin protein which specifically binds to an antigen. **Polyclonal** antibodies are produced in an intact animal whereas **monoclonal** antibodies are derived from a single clone of cells and are therefore specific for a single antigen.

Antigen: a foreign substance (usually a protein) which when administered to an animal will trigger the production of antibodies as part of the immune response.

Antiserum: antibody-containing serum from an animal after exposure to a specific antigen.

Atomic excitation: the jump of an electron to a higher (empty) orbital.

Atomic number: the number of protons in the nucleus of an atom, which is equal to the number of electrons.

Autoradiography: the detection of radioactivity using photographic film emulsions.

Beta (β) particle: a negatively or positively charged electron emitted from the nucleus of a radioactive atom.

Becquerel (Bq): the SI unit of radioactivity; defined as one disintegration per second (d.p.s.).

Blotting: the transfer of macromolecules from one flat surface (usually a gel) to another flat surface (usually a membrane).

Channel: the 'window' of an electronic pulse height analyzer programmed to accept electrical pulses of defined upper and lower energy levels by means of a discriminator.

Counting efficiency: the ratio of the observed counts per minute (c.p.m., or c.p.s.) to the absolute disintegrations per minute (d.p.m., or d.p.s.). Counting efficiency is usually expressed as a percentage (i.e. c.p.m./d.p.m. $\times$ 100).

Curie (Ci): a former unit of radioactivity; defined as 3.7×10^{10} d.p.s.

Cyclotron: a machine for accelerating positively charged particles to initiate nuclear reactions.

Decay: the disintegration of an unstable nucleus by the spontaneous emission of charged particles and/or photons.

Discriminator: an electronic device for allowing electrical pulses of a defined pulse height range to pass on to the next electronic unit.

Dissociation constant (K_D): the concentration of ligand required to give half-maximal receptor binding. The lower the K_D, the higher the affinity of the receptor for the ligand.

Dose-equivalence: the absorbed dose of radiation multiplied by the quality factor.

Electron: a stable fundamental particle of negative charge weighing approximately 1/1800th that of a proton or neutron.

Electron capture (EC): a mode of radioactive decay of a neutron-deficient radioisotope in which an orbital electron is captured by a proton in the nucleus with the emission of characteristic X-rays.

Electron volt (eV): the energy acquired by a unit (electronic) charge which has been accelerated through a potential of 1 volt.

Fluorography: a sensitive form of autoradiography incorporating an organic scintillator in the sample.

Gamma (γ) ray: a form of radioactive decay involving the emission of a discrete quantity of short wavelength electromagnetic radiation without charge or mass.

Gray (Gy): the SI unit of absorbed radiation; defined as 1 joule per kg.

Half-life ($t_{1/2}$). Radioactive: the time taken for a radioactive source to lose half of its activity by decay. **Biological**: the time taken for an organism to remove half of a compound by metabolism and/or excretion.

Hybridization: the detection of specific nucleotide sequences in DNA or RNA by complementary probes.

Imaging: recording the distribution of radioactivity in the body with the use of a sodium iodide detector.

Internal conversion (IC): a mode of radioactive decay in which a γ-ray, after emerging from the nucleus, transfers its energy to an electron in an orbital of its own atom, which is then emitted.

Internal transition (IT): a mode of radioactive decay in which a daughter nuclide is radioactive and itself decays with its own characteristic half-life by emission of γ-rays.

Ionization: the removal of an electron infinitely far from an atom to give an ion pair (i.e. a positively charged ion and an electron).

Ionizing radiations: any radiations which can ionize and excite the molecules and atoms which absorb them.

Isotopes: a series of nuclides with the same number of protons (i.e. the same element) but with differing numbers of neutrons.

Isotopic abundance: the percentage of atoms in the form of one isotope of an element in nature.

Ligand: any substance which binds specifically to a receptor.

Mass number (A): the sum of the number of protons (Z) and the number of neutrons (N) in an atom: $A = Z + N$.

Metastable state: after radioactive decay, the daughter radionuclide is left in an excited state and emits γ-radiation in order to reach the ground state. This normally occurs virtually instantaneously, but in a few cases the radionuclide has an excited state with a relatively long half-life (minutes to hours). Metastable states are designated by a superscript 'm' (e.g. ^{99m}Tc).

Negatron: a negatively charged electron emitted from the nucleus during some modes of radioactive deacy; also known as a β^--particle.

Neutrino: a neutral particle of zero (or very small) rest mass emitted during some modes of radioactive decay.

Neutron: a fundamental nuclear particle with a slightly greater mass than a proton, but no electrical charge.

Nucleon: a collective term applied to any particle (species) in the nucleus.

Nuclide: a general term describing any nucleus (stable or unstable) in terms of its atomic number and mass (i.e. an atomic species).

Quality factor (QF): a factor describing the relative ability of different radiations to cause damage to biological systems.

Quenching: any process which reduces the efficiency of measuring radioactivity.

Photon: a quantity of light or electromagnetic radiation.

Positron: a positively charged electron emitted from the nucleus during some modes of radioactive decay; also known as a β^+-particle.

Proton: a fundamental nuclear particle with nearly the same mass as a neutron and a positive electrical charge numerically equal to that of an electron (the nucleus of a hydrogen atom).

Radioactivity: the property of a material (atoms) to emit ionizing radiations.

Radioantigen: a radiolabeled antigen (usually with ^{125}I).

Radioisotope: an isotope of an element whose nucleus is radioactive.

Radioligand: a ligand containing at least one of its atoms as a radioisotope.

Radionuclide: an atomic species (nuclide) whose nucleus is radioactive.

Radiopharmaceutical: a compound, used in humans for either diagnosis or therapy, in which at least one of its constituent atoms is a radioisotope.

Receptor: a complex macromolecule (usually a glycoprotein) which specifically binds a messenger molecule (e.g. hormone, neurotransmitter).

Sievert (Sv): the SI unit of dose-equivalence; defined as Gy × quality factor.

Scintillator (scintillant): a chemical which emits quanta of light when excited by ionizing radiation from a radioisotope.

Specific activity: the amount of radioactivity per unit weight, for example, MBq/mg or MBq/μmol.

Tracer: a term loosely applied to describe a radioisotope introduced into a chemical or biological system to follow the path normally taken by the stable isotope.

X-ray: extra-nuclear electromagnetic radiation without charge or mass produced when fast electrons interact with a target.

Index